Coatings for High-Temperature Structural Materials

Trends and Opportunities

Committee of Coatings for High-Temperature Structural Materials
National Materials Advisory Board
Commission on Engineering and Technical Systems
National Research Council

NATIONAL ACADEMY PRESS
Washington, D.C. 1996

NOTICE: The project that is the subject of this report was approved by the Governing Board of the National Research Council, whose members are drawn from the councils of the National Academy of Sciences, the National Academy of Engineering, and the Institute of Medicine. The members of the committee responsible for the report were chosen for their special competencies and with regard for appropriate balance.

This report has been reviewed by a group other than the authors according to procedures approved by a Report Review Committee consisting of members of the National Academy of Sciences, the National Academy of Engineering, and the Institute of Medicine.

The National Academy of Sciences is a private, nonprofit, self-perpetuating society of distinguished scholars engaged in scientific and engineering research, dedicated to the furtherance of science and technology and to their use for the general welfare. Upon the authority of the charter granted to it by the Congress in 1863, the Academy has a mandate that requires it to advise the federal government on scientific and technical matters. Dr. Bruce M. Alberts is president of the National Academy of Sciences.

The National Academy of Engineering was established in 1964, under the charter of the National Academy of Sciences, as a parallel organization of outstanding engineers. It is autonomous in its administration and in the selection of its members, sharing with the National Academy of Sciences the responsibility for advising the federal government. The National Academy of Engineering also sponsors engineering programs aimed at meeting national needs, encourages education and research, and recognizes the superior achievements of engineers. Dr. Harold Liebowitz is president of the National Academy of Engineering.

The Institute of Medicine was established in 1970 by the National Academy of Sciences to secure the services of eminent members of appropriate professions in the examination of policy matters pertaining to the health of the public. The Institute acts under the responsibility given to the National Academy of Sciences by its congressional charter to be an adviser to the federal government and, upon its own initiative, to identify issues of medical care, research, and education. Dr. Kenneth I. Shine is president of the Institute of Medicine.

The National Research Council was organized by the National Academy of Sciences in 1916 to associate the broad community of science and technology with the Academy's purposes of furthering knowledge and advising the federal government. Functioning in accordance with general policies determined by the Academy, the Council has become the principal operating agency of both the National Academy of Sciences and the National Academy of Engineering in providing services to the government, the public, and the scientific and engineering communities. The Council is administered jointly by both Academies and the Institute of Medicine. Dr. Bruce M. Alberts and Dr. Harold Liebowitz are chairman and vice chairman, respectively, of the National Research Council.

This study by the National Materials Advisory Board was conducted under Contract No. MDA-972-92-C-0028.

This report is available from the Defense Technical Information Center, Cameron Station, Alexandria, VA 22304-6145.

Library of Congress Catalog Card Number 96-68712

International Standard Book Number 0-309-05381-1

Available in limited supply from:
National Materials Advisory Board
2101 Constitution Avenue, NW
HA-262
Washington, D.C. 20418
202-334-3505

Additional copies are available for sale from:
National Academy Press
2101 Constitution Avenue, NW
Box 285
Washington, D.C. 20055
800-624-6242 or 202-334-3313 (in the Washington metropolitan area)

Copyright 1996 by the National Academy of Sciences. All rights reserved.

Printed in the United States of America.

Abstract

This report assesses coatings materials and processes for gas-turbine blades and vanes; determines potential applications of coatings in high-temperature environments; identifies needs for improved coatings for performance enhancements, design considerations, and fabrication processes; assesses durability of advanced coating systems in potential service environments; and discusses required inspection, repair, and maintenance methods. Promising areas for research and development of materials and processes for improved coating systems and approaches for increased standardization of coatings are identified, with an emphasis on materials and processes with the potential for either improving performance, quality, or reproducibility or significantly reducing manufacturing costs.

ABSTRACT

COMMITTEE ON COATINGS FOR HIGH-TEMPERATURE STRUCTURAL MATERIALS

ROBERT V. HILLERY (chair), GE Aircraft Engines, Cincinnati, Ohio
NEIL BARTLETT, University of California, Berkeley
HENRY L. BERNSTEIN, Southwest Research Institute, San Antonio, Texas
ROBERT F. DAVIS, North Carolina State University, Raleigh
HERBERT HERMAN, State University of New York, Stony Brook, New York
LULU L. HSU, Solar Turbines, San Diego, California
WEN L. HSU, Sandia National Laboratories, Livermore, California
JOHN C. MURPHY, Johns Hopkins University, Laurel, Maryland
ROBERT A. RAPP, Ohio State University, Columbus
JEFFERY S. SMITH, Howmet Corporation, Whitehall, Michigan
JOHN STRINGER, Electric Power Research Institute, Palo Alto, California

National Materials Advisory Board Staff

ROBERT M. EHRENREICH, Senior Program Officer
CHARLIE HACH, Program Officer
JACK HUGHES, Research Associate
AIDA C. NEEL, Senior Project Assistant
ROBERT E. SCHAFRIK, NMAB Director
ROBERT SPRAGUE, Consultant
JILL WILSON, Program Officer

Technical Advisors

WILLIAM J. BRINDLEY, NASA Lewis Research Center, Cleveland, Ohio
STANLEY J. DAPKUNAS, National Institute of Standards and Technology, Gaithersburg, Maryland

Government Liasion Representatives

WILLIAM BARKER, ARPA, Arlington, Virginia
NORMAN GEYER, Wright-Patterson Air Force Base, Ohio
DAWN MIGLIACCI, Naval Air Warfare Center, Trenton, New Jersey
WILLIAM PARKS, U.S. Department of Energy, Washington, D.C.
ROBERT R. REEBER, Army Research Office, Research Triangle Park, North Carolina

NATIONAL MATERIALS ADVISORY BOARD

ROBERT A. LAUDISE (chair), Lucent Technologies, Inc., Murray Hill, New Jersey
G.J. (REZA) ABBASCHIAN, University of Florida, Gainesville
JAN D. ACHENBACH, Northwestern University, Evanston, Illinois
MICHAEL I. BASKES, Sandia-Livermore National Laboratory, Livermore, California
I. MELVIN BERNSTEIN, Tufts University, Medford, Massachusetts
JOHN V. BUSCH, IBIS Associates, Inc., Wellesley, Massachusetts
HARRY E. COOK, University of Illinois, Urbana
EDWARD C. DOWLING, Cyprus AMAX Minerals Company, Englewood, Colorado
ROBERT EAGAN, Sandia National Laboratories, Albuquerque, New Mexico
ANTHONY G. EVANS, Harvard University, Cambridge, Massachusetts
CAROLYN HANSSON, University of Waterloo, Waterloo, Ontario, Canada
MICHAEL JAFFE, Hoechst Celanese Research Division, Summit, New Jersey
LIONEL C. KIMERLING, Massachusetts Institute of Technology, Cambridge
RICHARD S. MULLER, University of California, Berkeley
ELSA REICHMANIS, Lucent Technologies, Inc., Murray Hill, New Jersey
EDGAR A. STARKE, University of Virginia, Charlottesville
KATHLEEN C. TAYLOR, General Motors Corporation, Warren, Michigan
JAMES WAGNER, The Johns Hopkins University, Baltimore, Maryland
JOSEPH WIRTH, Raychem Corporation, Menlo Park, California
ROBERT E. SCHAFRIK, Director

Acknowledgments

A great deal of work has gone into this study from its inception to the finished product. The committee is grateful for all the help it has received and expresses thanks to everyone who has participated. Without the patience and support provided by many individuals and organizations, this report could never have been completed.

The committee is grateful to those people who took time to brief the committee on the latest developments in coatings for high-temperature structural materials. The information and ideas from these briefings were essential to the study. Briefings were presented by Steve Balsone, Wright-Patterson Air Force Base; Andy Culbertson, Naval Air Warfare Center Trenton; Jack Devan, Oak Ridge National Laboratory; Dick Novak, Engineered Coatings; Randy Sands, Naval Air Warfare Center; Fred Soechting, Pratt & Whitney; Joseph Stephens, NASA Lewis Research Center; and Sharon Vukelich, Wright-Patterson Air Force Base.

The staff of the Southwest Research Institute (SWRI) did an excellent job of hosting one of the committee meetings, and the committee thanks Martin Goland, Henry Bernstein, and the SWRI staff for providing a tour of the facilities and accommodating the many needs of the meeting.

In particular, the committee thanks William Brindley of the NASA Lewis Research Center for making information available to the staff. The committee also thanks Stanley Dapkunas of the National Institute of Standards and Technology. Their support as technical advisors to the committee was invaluable.

The government liaisons who served this committee were also of enormous value. The committee thanks William Barker of ARPA, Dawn Migliacci of the Naval Air Warfare Center, William Parks of the U.S. Department of Energy, and Robert Reeber of the Army Research Office.

The chair of the committee thanks the members for their dedication and patience during the course of this study. This report could never have been completed without the diligence and goodwill of the members.

The committee thanks the staff of the National Materials Advisory Board, almost all of whom seemed to have been involved in the study at one time or another. Three program officers helped guide the study. In particular, the committee thanks Robert Ehrenreich for finishing the study. Tom Munns initiated the project and deserves thanks. Jill Wilson's support and guidance during the course of the study is greatly appreciated by the committee, as is Robert Sprague's, whose ideas as a consultant helped shape the report. Charles Hach and Jack Hughes were invaluable in the latter stages of the report. The committee also thanks Robert Schafrik for his support and direction along the way. Finally, the committee gratefully acknowledges the support of Aida C. Neel, senior project assistant.

ACKNOWLEDGMENTS..viii

Preface

The U.S. Department of Defense and the National Aeronautics and Space Administration requested that the National Research Council conduct a study and provide recommendations on future research and development needs for high-temperature coatings systems. This report represents the work of the Committee on Coatings for High-Temperature Structural Materials, established by the National Research Council for this purpose.

Performance improvements in high-temperature mechanical systems have resulted in increasingly severe operating environments for high-temperature structural materials, particularly in gas turbines. This has sparked an increased demand for more reliable coatings that possess predictable failure mechanisms, improve the performance of structural materials, and extend the operating range of applications. With this background, the following objectives were outlined for this study:

- assess the state of the art of coatings materials and processes
- identify potential applications for coatings in high-temperature environments
- identify needs for improved coatings for performance enhancements, design considerations, and fabrication processes
- assess durability of advanced coating systems in expected service environments
- identify required inspection, repair, and maintenance methods
- recommend promising areas for materials and process research and development for improved coating systems and identify approaches to increased coating standardization

To address these objectives, the committee considered (1) propulsion systems for commercial and military aircraft and their marine and industrial derivatives and (2) land-based turbines for power generation and mechanical drives (excluding automotive, diesel engines, and space applications). The committee directed its efforts toward the hot section (combustor and turbine) of these gas-turbine systems, because this represents the most significant materials and coating challenges. To focus the study further, the committee considered a wide range of application and technology areas that might be covered under this broad charter and identified those that would be reviewed in detail. The intent was to concentrate on the materials systems and degradation modes in the hottest section of the identified power-generation systems and to consider the technology and application implications for coatings systems under these most severe conditions. A primary focus was on the needs for advanced machines under development by the U.S. Department of Energy, the U.S. Department of Defense, and the National Aeronautics and Space Administration sponsorship. This deliberation resulted in the list shown in table P-1, which defines the study focus, the areas not considered, and the areas that were only referenced by association with the primary focus.

Through a series of briefings from industry and government experts, the committee reviewed current coating systems, newly developed coating systems, and their implementation in products over the next five to eight years. To evaluate coating needs beyond this time frame, the committee reviewed the substrate materials (e.g., ceramics and intermetallics) being considered for future engine designs. The committee recognized that defining needs for many future systems would currently lack clarity, but a need was perceived to anticipate any fundamental changes that may demand longer-range research, process development, or manufacturing innovations.

This report reviews the state of the art for coating systems based on the following approach. First, the application needs were identified and a description of the domain of use was developed. Second, the environment that currently exists and the substrate materials that are now used in the hot section of gas-turbine engines were examined. This, in turn, led to a more complete definition of the coatings systems required. Third, the application processes, the industrial base, and the repair and overhaul requirements were discussed and the support capabilities (e.g., modeling, testing, and nondestructive evaluation) were assessed. This review provided a baseline for discussion of future trends and indicated how U.S. industry, government, and academia are planning to address the requirements of advanced propulsion systems.

To determine materials and coatings needs, advanced systems were assessed. The assessments on these advanced systems were obtained through presentations and information provided by the program managers for three major

TABLE P-1 Committee Focus

Focus	Associated/Referenced	Excluded
Industrial, marine, power generation, aircraft engine	Land-based systems	Auto, diesel engines, space
Structural materials superalloys (Ni, Co) intermetallics composites (IMC, CMC) monolithic ceramics	Refractories	Titanium alloys
700°C + temperatures (to capture Type II hot corrosion)		
Oxidation, corrosion, erosion	Seal systems degradation caused by the service environment	Tribological wear
Gas-path coatings and clearance coatings	Gas-path seals	
Combustor, transition piece, high-pressure turbine, power turbine	Off-line combustion	Compressor, fan
Diffusion coatings, overlays, TBC, surface modifications	Functionally graded materials, claddings, vitreous coatings	Low observable coatings, fiber coatings
Thermal spray, CVD, PVD, advanced processes	Plating, C/S/N/O-resistant coatings	
Operating environment (air, fuel, water, particulates)	Combustion, emissions	
Environmental impact: manufacturing, service, overhaul		
Repair considerations	Lower-temperature coatings affected (e.g., impact-resistant coatings)	Repairs involving brazing, welding, etc.
Nondestructive evaluation (NDE)		
Standards and standardization		
Databases, modeling, engine condition sensors	Controls, intelligent processing of materials	
Systems design (coating/substrate integration): advanced concepts	Material systems that might reduce or eliminate need for high-temperature coatings (e.g., Lamalloy)	
Customer-DOD, DOE, NASA, original equipment manufacturers, suppliers	Airframe manufacturers, airlines, utilities	

TABLE P-2 Primary Performance Goals for Advanced Engine Systems

Advanced Engine System	Requirements
Integrated High-Performance Turbine Engine Technology (IHPTET)	Cold section (fan and compressor): 2-3 times specific strength for materials 650-1000°C operating temperature
	Hot section (the focus of this report; includes the combustor, turbine, augmenter, and nozzle subsytems): 3-5 times specific strength for materials 1650-2200°C with advanced cooling 1550°C uncooled
	Nonstructural (bearing and lubes) up to 825°C
High-Speed Civil Transport (HSCT)	Range: 2 times greater than the Concorde Payload: 3 times greater than the Concorde Economics: 8 times greater than the Concorde Environmental emissions: 8 times lower than the Concorde Noise: 3 times quieter than the Concorde
Advanced Turbine Systems (ATS)	High efficiency, clean gas-turbine systems initially based on natural gas; adaptable to coal- or biomass-derived fuels
	Power generation: >60% system efficiency
	Industrial systems: >15% improvement
	Environmental: 8 ppm NO_X emissions; CO and HC < 20 ppm
	Cost competitive: 10% reduction in busbar cost of electricity

government-sponsored materials efforts representing future military, commercial, and power-generation needs:

- Integrated High-Performance Turbine Engine Technology—advanced military systems
- High-Speed Civil Transport—aimed at the advanced supersonic commercial market
- Advanced Turbine Systems—advanced utility and industrial power generation

Table P-2 summarizes the primary performance goals for each of these advanced systems. For each case, the committee obtained information on mission profile; systems needs; and specifics on time, temperature, and environmental requirements for materials in the propulsion system. This provided a perspective on what were the ultimate materials needs for propulsion systems reaching maturity early in the next century. In all cases, the goals underscored the demand for materials that can withstand significantly higher operating temperatures and service life than today's state-of-the-art devices. The reviews also provided information on significant changes that might be required as a result of new regulatory requirements, such as those that might stipulate permissible emissions from future propulsion systems. In addition, these reviews showed those areas common to aircraft engines and power-generation machines and changes to this commonality that might be demanded by derivative machines as, for example, the increased use of air to combat NO_X in the combustor of land-based electric utility turbines. Some fundamental differences exist between aircraft and land-based systems that might cause a divergence in materials (and coatings) technologies. For instance, land-based systems are less affected by weight and can be supplemented with auxiliary systems, such as air supplies, steam supplies, or heat exchangers. The potential use of alternative fuels in nonaircraft systems might be another divergence affecting coatings type. Both aircraft and land-based gas turbines are moving toward higher temperatures and longer service-time requirements; this trend is causing increased emphasis on coatings needs. Finally, the committee heard presentations on the design requirements for coating systems and the engineered materials efforts that may have a bearing on the development and application of advanced coating systems.

In reviewing these briefings, the committee considered the following key questions:

- Can the goals for the advanced systems be achieved simply by an evolution from today's materials?
- Are programs and efforts in place to address the key potential barriers?

- Are there additional recommendations that can be made to enhance the chances for success in any of the key areas?

The committee then considered future opportunities for developing improved coatings by virtue of evolutionary development and by way of innovative concepts. The committee developed several innovative concepts for advanced coating systems and suggested how a wide variety of ideas could be integrated into coatings development and application advances. The members of the committee continually posed several key questions during their considerations of innovative approaches:

- Are there concepts that have not been explored or that should be re-evaluated in light of recent knowledge?
- Are there ideas or knowledge bases in other industries that can revolutionize thinking in the gas-turbine engine coating industry?
- Are there ideas that can shorten the development cycle of existing efforts and thereby enhance the chance for success?
- What are the latest developments in modeling, intelligent process manufacturing, and smart materials, and can these technologies be focused in the coatings area?
- Are there new testing techniques that would be required and will new industry standards and procedures have to be defined?

Finally, the committee provides its conclusions and recommendations for the future as well as a bibliography of cited references and available texts.

Robert V. Hillery, chair
Committee on Coatings for High- Temperature Structural Materials

Contents

	EXECUTIVE SUMMARY	1
1	INTRODUCTION	8
2	APPLICATION NEEDS AND TRENDS	10
	Application Needs	10
	Current Component Design and Base Materials	13
	Advanced Materials for Engine Components	13
3	MATERIALS AND PROCESSES	19
	Coatings for High-Temperature Structures	19
	Coating Processes	22
	Coating Process Control	24
	Summary	24
4	FAILURE MODES	26
	Degradation Mechanisms of Structural Materials	26
	Degradation Mechanisms of Coatings	26
	A Case Study: Degradation of Thermal Barrier Coatings	30
	Research Opportunities	32
5	ENGINEERING CONSIDERATIONS	34
	Compatibility of Coatings with Structural Materials	34
	Component Coatability	36
	Other Engineering Considerations	36
	Concurrent Coating Development	38
6	REFURBISHMENT OF COATED STRUCTURE	39
	Factors Affecting Component Life	39
	Repair of High-Temperature Coatings	40
	Standard Designations for Coatings	41
	Nondestructive Evaluation	41
7	NEAR-TERM TRENDS AND OPPORTUNITIES	43
	Thermal Barrier Coating Development	43
	Coating Processes	44
8	LONG-TERM OPPORTUNITIES AND INNOVATIVE SYSTEMS	46
	Innovative Coating Architectures	46
	Other Innovative Concepts	48

	REFERENCES	51
	APPENDICES	
A	TESTING AND STANDARDS	57
B	RADIATION TRANSPORT IN THERMAL BARRIER COATINGS	65
C	SURVEY OF NONDESTRUCTIVE EVALUATION METHODS	67
D	MODELING OF COATING DEGRADATION	72
E	MANUFACTURING TECHNOLOGIES OF COATING PROCESSES	78
F	EXAMPLE OF A COATING DESIGNATION SYSTEM	83
G	BIOGRAPHICAL SKETCHES OF COMMITTEE MEMBERS	84

List of Figures and Tables

FIGURES

1-1	Typical ranges of a coating property compared with advanced engine requirements for a property	9
2-1	Cross section of a typical modern gas-turbine engine	10
2-2	Overview of transportation gas-turbine use	11
2-3	Overview of power-generation and mechanical-drive gas-turbine use	11
2-4	Approximate increases in firing temperature capabilities from 1980 to 2010	12
2-5	Turbine blade cooling methods	15
3-1	Coating compositions as related to oxidation and corrosion resistance	21
3-2	Schematic showing the benefit of development and deployment of manufacturing process control for high-temperature coatings	25
4-1	Types of high-temperature attack for metallic coatings (aluminide, chromide, MCrAlY, etc.) on nickel-base superalloys with approximate temperature regimes and severity of attack	28
4-2	Micrograph of service-exposed CoCrAlY overlay coatings showing internal oxidation of coating and base metal	29
4-3	Comparison of the microstructure of EB-PVD and plasma-sprayed TBCs	30
4-4	Photomicrograph of a plasma-sprayed TBC	31
4-5	Photomicrographs of EB-PVD TBCs before and after failure	32
5-1	Tradeoffs between first cost and operating cost	37

TABLES

P-1	Committee Focus	x
P-2	Primary Performance Goals for Advanced Engine Systems,	xi
2-1	Typical Duty Cycles for Various Gas-Turbine Engines	12
2-2	Nickel-and Cobalt-Base Alloys Used in Land-Based and Aircraft Gas Turbines	14
2-3	Properties of Ceramic Materials that are Candidates for Hot-Section Use	16
2-4	Selected Properties of High-Temperature-Capable Intermetallics Compared to a Conventionally Cast and a Single-Crystal Superalloy	17
2-5	Properties of Refractory Metals and an Alloy Compared to B 1900 Nickel-Base Superalloy	18
3-1	Coating Functions and Coating Materials Characteristics	19
3-2	Types of Coatings Used in Hot-Section Components	20
3-3	Generic Information on Coating Types Used in Superalloy Hot-Section Components	21
3-4	Summary of the Benefits and Limitations of the Atomistic and Particulate Deposition Methods	22
4-1	Environmentally Induced High-Temperature Structural Material Failure Modes	27
4-2	Environmentally Induced High-Temperature Coating Failure Modes	28
6-1	Survey of Nondestructive Evaluation Techniques	42

Acronyms

AISI	American Iron and Steel Institute
ASM	American Society for Materials
ASME	American Society of Mechanical Engineers
ASTM	American Society for Testing and Materials
ATS	Advanced Turbine Systems
CTE	coefficient of thermal expansion
CVD	chemical vapor deposition
DOD	U.S. Department of Defense
DOE	U.S. Department of Energy
EB-PVD	electron-beam physical vapor deposition
FGM	functionally graded materials
HSCT	High-Speed Civil Transport
HVAF	high-velocity air fuel
HVOF	high-velocity oxy fuel
IHPTET	Integrated High-Performance Turbine Engine Technology program
ISO	International Standards Organization
LPPS	low-pressure plasma spraying
NASA	National Aeronautics and Space Administration
NDE	nondestructive evaluation
PVD	physical vapor deposition
SNECMA	Société National d'Etude et de Construcion de Moteurs d'Aviation
STEP	Standard for the Exchange of Product
SVPA	SNECMA vapor phase aluminizing
TBC	thermal barrier coating

Executive Summary

Traditionally, coatings and substrates have developed independently. Coatings have also traditionally done an excellent job of doing what they were designed to do: prolong the life of turbine engines by protecting component parts from oxidation and corrosion, erosion by particulate debris, and other potential hazards. Engineers now face a challenge, however. With new technologies creating a broad range of heat-resistant materials, turbines now operate at temperatures that are significantly higher than a decade ago. The new demands on turbine coatings and substrates make it imperative that the two be designed *interdependently;* each must go hand-in-hand into the regime of ever-increasing temperatures. In this harsh environment, a failure in one quickly leads to a failure in the other. Indeed, in some proposed designs, the coating and substrate form a continuum, literally blurring the boundary between the surface deposit and the material it coats.

In future years, turbine engines will have the potential to reach new heights of efficiency and service life. But to keep pace, coating technologists will have to continue moving away from the traditional way of designing coatings. The bottom line is that coatings must be integrated into the total component design taking into full consideration the alloy composition, casting process, and cooling scheme.

The efficiency of gas turbines, whether for industrial power generation, marine applications, or aircraft propulsion, has steadily improved for years. These advances have come about, in large part, because the means have been found to operate the gas-generator portion of the engine at increasingly higher temperatures. The need for greater performance from advanced turbine engines will continue, requiring even higher operating efficiencies, longer operating lifetimes, and reduced emissions. A large share of these improved operating efficiencies will result from still higher operating temperatures. Better engine durability would normally require lower operating temperatures, more cooling of the hot structure, or structural materials possessing inherently greater temperature performance. Since the first two options cause a penalty in operating efficiency, the last approach is preferred. Achieving greater temperature performance has made imperative the use of surface protection to extend component life and the concurrent development of the advanced structural materials and the coatings that protect the structure from environmental degradation.

This report assesses the state of the art of turbine coatings, identifies applications for coated high-temperature structures, identifies needs for improved coating technologies, assesses durability of coatings in expected service environments, identifies coating life-cycle considerations, suggests innovative directions for coating systems, and presents recommendations for coating technologies. *The report concludes that coatings have become an enabling technology for advanced engines; the development of coatings and their processes must keep pace with the broader materials and systems requirements.*

TRENDS

High-Temperature Coatings Design

In the past, high-temperature coatings were selected predominantly after the component design was finalized. Current designs require that the substrate (typically a nickel-base superalloy) have sufficient inherent resistance to the degradation mechanisms to prevent catastrophic reduction in service lifetime in the event of coating failure. Since the materials considered for future substrates may possess less inherent environmental resistance at higher temperatures, the importance of coatings in achieving performance will continue to grow. In future turbine designs, coatings will be increasingly viewed as an integral portion of the design process to meet the high demands for system performance.

High-Temperature Coating Types

Although many types of high-temperature coatings are currently in use, they generally fall into one of three types: aluminide, chromide, and MCrAlY.[1] The family of coatings that insulate the substrate from the heat of the gas path (i.e., thermal barrier coatings [TBCs]) is increasing in importance as they begin to be used for performance benefits. TBCs are

---

[1] MCrAlY is a type of metallic coating in which M is a metal, usually cobalt, nickel, or a combination of the two; Cr is chromium; Al is aluminum; and Y is yttrium.

ceramic coatings (e.g., partially stabilized zirconia) that are applied to an oxidation-resistant bondcoat, typically a MCrAlY or aluminide.

Processes for Applying Coatings

A wide variety of processes are used to apply coatings, although they rely on one of three general methods: physical vapor deposition, chemical vapor deposition, and thermal spray. These processes deposit a wide range of coatings between the extremes of diffusion coatings (i.e., the deposited elements are interdiffused with the substrate during the coating process) and overlay coatings (i.e., the deposited elements have limited interdiffusion with the substrate). Diffusion coatings are well bonded to the substrate but have limited compositional flexibility; their usefulness is strongly dependent on substrate chemistry. Overlay coatings are typically well bonded and have broad compositional flexibility; however, they are more expensive and thicker than diffusion coatings. TBCs are overlay coatings and as such can be deposited on a variety of substrates. The main difficulty with TBCs is that the abrupt change in composition and properties at the interface tends to promote ceramic layer spallation.

Electron-beam physical vapor deposition is often favored over plasma deposition for TBCs on turbine airfoils since it applies a smooth surface of better aerodynamic quality with less interference to cooling holes. However, the widely used plasma-spray process has benefits, including a lower application cost, an ability to coat a greater diversity of components with a wider composition range, and a large installed equipment base.

Coating developers must not only find a suitable coating for an application but must also develop the necessary application processes with on-line control so that the resultant composition and microstructure of the coating is highly reproducible and within the performance limits needed for the service requirements. Developing the relationship of the process-to-product performance must also be a priority, near-term endeavor for advanced coating systems.

Degradation Modes

A primary consideration in selecting a coating system is determining if it provides adequate protection against the active, in-service, environmentally induced degradation mechanism(s) experienced by the component. These degradation modes are a function of the operating conditions and the component base materials. The degradation modes common to superalloy hot-section components include—to varying degrees—low-cycle thermomechanical fatigue, foreign object damage, high-cycle fatigue, high-temperature oxidation, hot corrosion, and creep.

Because of the use of thin walls and compositional design for highest strength, aircraft turbine blades with internal cooling passages have historically had insufficient high-temperature oxidation resistance to meet required lifetimes without the use of a coating. Coatings have been used in these circumstances to extend overhaul limits and useful life of the component. Although the latest generation of single-crystal blades has excellent oxidation resistance compared with conventionally cast industrial engine blades and aircraft gas-turbine blades with moderate to high chromium contents, the blades have less tolerance for hot corrosion once the coating has been breached. Industrial gas-turbine blades, which use thick walls and lower-strength alloys with higher corrosion resistance, generally have significant service life after the coating is breached.

During service, coatings degrade at two fronts: the coating/gas-path interface and the coating/substrate interface. Deterioration of the coating surface at the coating/gas-path interface is a consequence of environmental degradation mechanisms. Solid-state diffusion at the coating/substrate interface occurs at high temperatures, causing compositional changes at this internal interface that can compromise substrate properties and deplete the coating of critical species. In the worst case, interdiffusion leading to the precipitation of brittle phases can cause a severe loss of fatigue resistance.

Engineering Considerations

Given that a coating system is required and that one has been identified that provides environmental protection, six significant engineering factors must be evaluated.

1. Chemical (metallurgical) compatibility. The coating must be relatively stable with respect to the substrate material to avoid excessive interdiffusion and chemical reactions during the service lifetime. An unstable coating can lead to premature degradation of both the coating and the substrate through lower melting temperatures, lower creep resistance, embrittlement, etc.
2. Coating process compatibility. The coating material may be completely compatible with the component, but the coating process may not be compatible. This would usually occur when process conditions require high temperatures or special precoating surface treatments.
3. Mechanical compatibility. Coatings resistant to oxidation and corrosion maintain their protectiveness only if they remain adherent and free from through-thickness cracks. Important considerations include close match of the coefficient of thermal expansion (CTE) of the coating with the substrate,

strain accommodation mechanisms within the coating, coating cohesion, and coating adhesion. CTE match is the most important factor, closely followed by the need for strain tolerance in the coating.

4. Component coatability. The ability to deposit a coating on the required surface is a function of the geometry and size of the component, as well as the capability of the coating process. Accessibility of the surface is a consideration. For example, some processes are line-of-sight and thus cannot coat internal passages. Size of the component is important because some processes must be done inside an enclosed tank or reactor. The ability to apply a uniform coating must be evaluated, particularly at edges, inside corners, and for irregular part contours. The change in part dimensions and surface characteristics because of the coating must also be taken into account.
5. Contaminants in air and fuel (and water and steam for industrial turbines). Contaminants can combine in the hot section to produce corrosion, erosion, and deposition under certain temperature and pressure conditions; they contribute to accelerated degradation of high-temperature components. Limits on allowable concentrations must be established in order to assure the effectiveness of a coating system.
6. Turbine emission levels. Gas turbines can produce harmful emissions as part of the combustion process. As combustion technology has improved, emission levels have been reduced. These emissions include nitrogen oxides (NO and NO_2, commonly called NO_X), carbon monoxide (CO), unburned hydrocarbons, sulfur oxides (mainly SO_2 and SO_3), and particulate matter. Coatings affect emissions primarily by reducing the need for cooling air.

In addition to the factors pertaining to the selection of an appropriate coating system, the following general engineering considerations are also important:

- Available databases of coating and coated structure properties. Traditionally, engineering property data for high-temperature coatings are generated after the mechanical properties of the uncoated substrate have been well characterized. These data are generally specific to the application domain and process conditions and are usually proprietary. Long-term data (i.e., performance of coated structures for durations greater than 50,000 hours) is sparse and related to old technologies.
- Coating standardization. Generally, each component in the hot section of the engine has a particular coating system optimized for the prevailing conditions. Greater consideration should be given to optimizing a coating system for many components because of the wide variety of alloys and component systems.

Life-Cycle Factors

Hot-section structures are designed to operate at the highest possible temperatures and stresses in order to maximize performance. As a consequence, these structures continuously degrade during service. The rate at which this degradation occurs is crucial to the function of the component and, ultimately, to the performance and longevity of the gas turbine.

The role of the hot-section coating is to protect the substrate from the gas-path environment in order to meet performance objectives, as manifested in the time between overhauls or the designated service. Component refurbishment involves the economical and timely restoration of part integrity.

The types of repairs allowed to coated structures are dictated first by safety and reliability and second by economic benefit. Repairs therefore vary greatly depending on the type of component to be refurbished. Although the replacement of the coating is generally a small portion of the overall repair, it can be critical to meeting the intended life of the component after it is returned to service. The wide variety of coating systems and the lack of standard designations adds complexity to the logistical task of maintaining an engine's coated structure complex. This task will only become more difficult as advanced coatings find their way into service.

In the past, coatings had to be capable of being removed and reapplied. This may not continue to be a requirement for industrial turbines. If a new coating could allow higher-temperature operation (for increased efficiency), the savings in fuel costs could possibly outweigh the extra expense of purchasing new parts versus repairing old parts. The future trends for aircraft engine repairs will tend to parallel those of the industrial turbines with the further complication of thinner walls and more sophisticated cooling passages. Thinner walls in advanced components may preclude any stripping of the prior coating, potentially leading to a nonrepairable part, as is the case with many of the current turboprop and turboshaft high-pressure turbine blades.

NEAR-TERM OPPORTUNITIES

Concurrent Development of High-Temperature Coatings and Substrate Materials

Coating and substrate development are increasingly done concurrently because the coating and the substrate are becoming, from all the major life-cycle considerations (e.g. design, manufacturing, and product support), an integral entity. This concurrent development applies to the current generation of MCrAlY coatings and, notably, to the emerging TBC technologies.

The current generation of MCrAlY coatings, as well as the emerging TBC technologies, would benefit significantly from advances in process control. Both types of coatings are deposited by similar processes, and improvements will enhance the performance of both coating types. *Improved on-line control should be developed to ensure that the resultant behavior of the coated structure is highly reproducible and within the performance limits needed for the service requirements.*

Process control cannot be achieved, however, without understanding the relationship of the process to product performance. *Development of such knowledge must be a near-term priority for advanced coatings.* While engine tests do not necessarily provide data on individual processes, they are necessary to provide an overall qualification for a new coating system.

The similarities in coatings needs for power generation and aircraft engines are more significant than their differences. The view on traditional materials held that power-generation machines derive their benefits from the aircraft engine technology. As industry, government, and academic cooperation and consortia in materials development become more prevalent, and to the extent that the development of new coatings is done jointly among manufacturers and suppliers, there will be a move to even more similar coatings in the marketplace.

Oxidation and Hot-Corrosion-Resistant Coating Development

The demands on coatings have evolved since coatings were first applied to gas-turbine airfoils in the 1950s. Coatings have historically been developed to provide protection against oxidation and hot corrosion. Oxidation-resistant coatings typically are either aluminide coatings or overlay coatings with high aluminum activity that form an adherent alumina scale. Hot-corrosion-resistant coatings also rely on alumina as the protective scale, but in addition generally contain higher levels of chromium to ameliorate the effects of sulfur. *Incremental developments to improve the durability of oxidation and hot-corrosion-resistant coatings for current generation engines will be made by (1) chemistry modifications to both diffusion aluminide and overlay MCrAlY systems and (2) more stringent control of undesirable elements in both the substrate alloys and the coatings.*

The potential for higher-temperature use for superalloys is limited by their melting points. Thus, alternative component materials are being investigated to fill the higher-temperature roles. These materials generally fall into three classes: ceramics, intermetallics, and refractory metals. These materials significantly differ from superalloys in physical, chemical, and mechanical properties but will still depend on coatings to protect against environmental degradation. Many of these emerging materials, in their current form, are much less tolerant of flaws and failure in their coatings than the superalloys. Therefore, coating of these materials presents significant challenges.

Thermal Barrier Coating Development

TBCs reduce the severity of thermal transients and lower the substrate temperature, enhancing the thermal fatigue and creep capabilities of coated components. In addition, although TBCs do not provide significant reduction in oxygen transport to the substrate, the lower component temperature can lead to a reduction in oxidation and hot corrosion.

TBCs are finding increased application in overall component design. Over the past 25 years, cooling technology has contributed roughly 370°C (700°F) (from solid blades to advanced film cooling) in turbine temperature capability; further advances may be achieved with even more sophisticated cooling schemes. Superalloy material and processing advances (from equiaxed crystalline structure to third-generation single crystal) have added approximately 120°C (250°F). However, superalloys now operate in some applications at 90 percent of their melting point. TBCs have the potential to reduce substrate temperatures by 110°C (200°F) or more, even with current production methods.

Current knowledge of TBC durability and thermal performance is primarily in the form of empirical data. Surprisingly little is understood about TBCs in the critical areas of radiative heat transfer, thermal conductivity and emissivity, and fundamental physical and mechanical properties (i.e., fatigue, monotonic properties, or time-dependent properties). For example, until recently, the energy transport process in TBCs has been characterized by an effective thermal conductivity without regard to the relative contributions of radiation and true conduction.

Research is required to enhance understanding of TBC behavior (and thereby improve TBC performance) as well as to provide reliable information for quantitative modeling. To determine the relative importance of radiative and conductive transport in TBCs, a number of factors affecting these two thermal energy transport processes must be considered. *Attention should be given to understanding the mechanisms of energy transport in TBCs.* Knowledge of the relative importance of these mechanisms will guide research strategies aimed at reducing energy transport rates.

Reliability is a critical design factor that is in need of further development if TBCs are to be fully exploited to increase turbine efficiency. TBCs fail as a result of erosion, impact damage, interfacial oxidation of the bondcoat, or thermomechanical strain at the ceramic/metal interface. These factors and process variability combine to give current generation TBCs wide variability in service life. The lack of reliability, more than any other design factor, has slowed the introduction of these coatings for turbines. *Improved understanding of interfacial behavior is required to control coating properties and predict performance. A more compatible and oxidation-resistant bond between the TBC and either the metallic substrate or the bondcoat requires continued near-term emphasis.* The processes by which TBCs are currently applied, namely plasma spray and physical vapor deposition, will likely continue to be the major manufacturing methods.

Analytical Methods and Models

Coating producers and users need data and advanced analytical models. Examples include data on long-term thermodynamic and structural stability, generic process models, and life-prediction models. Appropriate combinations of methods must allow measurement, on the scale of the coating (which is from 1 to 30 mil), of specific materials properties such as adhesion, fracture toughness, thermal conductivity, and elastic modulus. The availability of such methods can provide the basis of standard test methods for coating assessment.

The coating industry needs precompetitive[2] research that, particularly for TBCs, identifies critical properties and the scale on which they are relevant. This research would provide the basis for developing standard procedures for measuring and comparing properties of coating systems as well as providing data required for use in performance models (e.g., methodologies for measurement of thermal conductivity, interfacial adhesion, and microstructural characterization).

Since future engines will rely heavily on coatings to protect hot-section components, accurate models will become essential to describe a number of coatings-related requirements, particularly:

- key attributes of the manufacturing process (e.g., microstructure, rate of coating deposition, cost, etc.)
- degradation modes (e.g., oxidation and corrosion)
- life-prediction and residual-life assessment

Process modeling is most important to coating manufacture. Degradation modeling is most important to coating design and development. Coating life and inherent substrate environmental resistance are key determinants in setting the intervals for engine inspection and overhaul. Few models exist in the public domain that address any of these needs.

Repair and Overhaul

Engineers currently rely heavily on visual inspection to assess the condition of coated structure. As a result, in-service condition monitoring and repair decisions focus on deterioration at the coating/gas-path interface. Improved nondestructive evaluation methods would provide information on when the coating has to be removed and on the extent of base-metal attack.

An important need for the repair of industrial gas-turbine components is industrywide repair specifications and regulation of the quality of repairs. The most effective method to achieve a consensus of all interested parties is unclear. Methods to make local repairs of coatings are needed, both during manufacture and operation. Also required are standard, industry-accepted methods to determine the durability and properties of refurbished coatings.

Aircraft engine repair needs parallel those for the industrial turbines, with the further complexity that these coating structure systems include more advanced designs and materials that tend to limit repair options. Incorporation of better models and data from condition-monitoring sensors will improve repair/replace decisions. Still unclear is the extent to which many of the advanced coatings, such as the TBCs, lend themselves to repair. Although some TBC overhaul is currently done, the extent to which TBC-coated components can be repaired and re-used has never been fully determined.

Although generic families of coatings exist, there is no standard system of designating or defining specific coatings within a family. Each manufacturer and vendor uses a unique nomenclature, a practice that causes confusion during the refurbishment of the coated components. *There are enough similar coatings in common use that a standard designation system would be practical and useful.*

Nondestructive Evaluation

Exploiting existing and advanced nondestructive evaluation (NDE) methods can aid significantly in developing and qualifying coating systems, improving process control during coating operations, and characterizing the integrity of coated structure during turbine engine manufacture, in-service condition monitoring, and repair and overhaul operations. Since each of these applications has specialized requirements, no single NDE method will likely serve all purposes. Development programs for advanced NDE methods should focus on supporting these key areas with the goal of bringing the new methods into practice. The highest priority for further NDE development should be for those noncontacting methods that can examine the interior structure of the coating system, such as the coating/substrate interface and base metal.

For aircraft engines in particular, the development of advanced NDE techniques and cost-benefit models will be

---

[2] Precompetitive research provides data in nonproprietary areas where competing companies can comfortably collaborate. An example is basic property data for commonly used materials.

essential for the assessment of components that are expected to be multiwall or thin-wall structures with multilayered coatings used as an integral part of the component design and manufacture.

LONG-TERM OPPORTUNITIES AND INNOVATIONS

Future generations of higher-performance aerospace turbine structures will require advanced materials, because inservice superalloys are approaching the upper limit of their inherent temperature capability. Candidate materials under consideration to replace superalloys include intermetallic compounds, monolithic and composite ceramics, and refractory alloys. In addition, advanced cooling concepts will result in processing modifications and more complex cooling paths to meet the demands of advanced component designs. Most advanced materials and design modifications will result in component structures with inherently less resistance to aggressive environmental attack than current superalloys, pushing the need for parallel development of improved coatings systems. *Incremental improvements to current coating technologies are unlikely to meet the goals of future-generation, higher-performance turbine engines. Innovative concepts are required.*

In the long term (i.e., beyond five to ten years), close integration of the coating and substrate material will be standard practice for key structural components. Control of properties at the substrate/coating interface will be critical. Future coatings will likely have graded compositions and multiple layers. They will be expected to operate at even higher temperatures and in steeper thermal gradients. The need for adherence and metallurgical stability completes the maze of major requirements. Coating developers must also meet the technical requirements in a cost-effective manner.

Innovative Coating Architectures

The structural motifs for hybrid coatings include multilayered materials, materials with ordered vertical structures such as channels, and materials with an intrinsic three-dimensional pattern such as dendrites or whiskers. In all cases, an important element that must be considered is the stability of such structures in the high-temperature environment of an operating engine. This stability includes the long-term ability of the material to maintain its initial mechanical properties and chemical and microstructural morphology. Research is needed on the following issues of stability:

- *Continuously graded coatings.* Graded coatings, such as functionally graded materials (FGM) and nanostructures, offer potential advances in coating performance. The need also exists for ceramic coatings that can withstand higher temperatures. Graded coatings may demand alternative materials as well as alternative application processes. *Most significant in this area would be a critical assessment of the use of FGMs as coatings and a definition of the influence of multilayer and nanostructure morphology on resulting properties.* However, the inherent inhomogeneity in the microscopic scale of these materials raises questions of high temperature stability that must be answered to establish the viability of these approaches. Advanced substrates such as composites and ceramics will possess coatability characteristics (e.g., diffusion rates and surface chemistry) very different from current superalloys. Novel concepts may be needed for coatings and surface treatments to protect these substrate systems.
- *Horizontally layered materials.* Layered materials offer a number of opportunities for advanced coating concepts. For example, the heat transfer may be reduced by making the coating a high-temperature optical multilayer interference filter that can reflect radiation or reduce conductivity.[3] If successful, such new TBC coatings would permit significantly higher turbine engine gas temperatures. Multilayers may also permit the matching of thermal expansion coefficients to minimize thermally induced internal stresses in the coating and at the coating/substrate interface, although this concept requires further examination using mechanical modeling and experimentation. Again, high-temperature stability is an issue to be resolved.
- *Interphase layer.* The crucial interface in any system designed to use an oxide as coating or bondcoat for protection in high-temperature environments is the *substrate-to-first-oxide layer.* If this interface could be replaced by an interphase layer, the composition of which varies continuously and gradually from the substrate metal (or ceramic) to the full oxide, spalling of the oxide layer from its substrate could be less likely to occur. Previous attempts to use a compositionally graded metal/ceramic interfacial layer have resulted in oxidative expansion that caused the oxidized graded layer to buckle the coating away from the substrate (Duvall and Ruckle, 1982; DeMasi-Marcin et al., 1989). There may be potential for other gradient schemes, however.
- *Vertically layered materials.* The use of vertical structure in the coating or substrate may also offer important advantages for improved thermal isolation of the substrate from the engine environment. For example, TBCs must meet the conflicting demands of adherence and low heat transfer between the coating and substrate. A

---

[3] Current thinking on nanolaminates is that reduced conductivity depends more on increasing phonon scattering than increasing reflectivity.

potential solution is to take advantage of the lithographic patterning technology that is widely used in the semiconductor industry to create high-aspect-ratio (high depth with narrow width) trenches that limit heat transport down to the substrate. Subsequent steps would fill in the trenches with a material that adheres poorly but has low thermal conductivity and then cap the entire coating. Such a channel structure could also provide a means to reduce the radiative heat load if the repetition distance between channels is commensurate with the infrared wavelength for which the unstructured coating is transparent. This condition would lead to diffraction of infrared radiation from the channel structure, effectively increasing the reflectance of the coating/substrate system.

- *Three-dimensional structured materials.* Another approach involves the formation of three-dimensional networks to provide improved mechanical stability and possibly provide resistance to heat conduction. For example, whiskers could be formed in situ through a phase transformation. If sintering, densification, and the desired amount of bonding between the matrix and whisker phases can be applied successfully to ceramic coatings, the processes could yield robust composite coatings that are strong, lightweight, highly dense, and resistant to severe environments.
- *Advanced processing.* Coatings processes developed in the electronics industry for the manufacture of semiconductor devices have potential applications for coatings for high-temperature structural materials. Electronics processing relies increasingly on in situ monitoring and process diagnostics (intelligent processing) to achieve nanoscale structural control and characterization. Such techniques might be adapted for turbine coatings in order to improve coating quality and increase the cost effectiveness of coating processes for the manufacture of functionally graded coatings. *The committee believes that intelligent materials processing will be required to achieve the reproducible process control necessary for manufacture of reliable coatings.* Intelligent materials processing requires the availability and integration of process models that describe the relationships of critical process parameters with sensors that measure these parameters and appropriate control systems. Development of all these elements is required.

Other Innovative Concepts

A variety of other approaches for improved coating/substrate systems are also possible. In most cases, there has been no proof of concept or even prior work for these ideas. They are presented as ideas to stimulate new research directions.

- Built-in sensors for condition monitoring. Microsensors could be embedded into coatings, such as TBCs, to monitor local temperature rises, changes in oxidation, and possibly incipient disbonding. These sensors would act as real-time monitors of the degradation of protective coatings and may serve to warn of imminent catastrophic failure. Other sensing technologies may use remote sensors rather than embedded sensors to achieve the same goal.
- Embedded microchannels within TBCs for cooling. Microdesigned coatings could include small cooling channels within the TBC to reduce further the surface temperature of the substrate. A cooling fluid, such as air or helium gas, could be forced along these channels to reduce the heat load transported to the substrate. Alternatively, a fluid (gas) that has poor thermal conductivity might be injected into the channels to act as an insulating sheath that limits heat conduction from the TBC.
- Coatings for refractory metals. Refractory metals are attractive potential substrates because of their high melting temperatures and high-temperature strength. Their susceptibility to catastrophic oxidation, however, is the main obstacle to their use in advanced turbine applications. Coatings that have an improved oxidation barrier offer improved performance of refractory metals. New coating materials may alloy electron-rich noble metals with electron-poor metals to form remarkably stable compounds with close-packed, or nearly close-packed, structures. These hybrid materials could be exploited to develop a coherent, tenacious coating.

1
Introduction

In future years, gas-turbine engines of all types will have the potential to reach new heights of efficiency and service life. But to keep pace, engineers will have to design coatings concurrently with the intended substrate. Indeed, for some proposed designs, the coating and substrate form a continuum, literally blurring the boundary between the surface deposit and the material it coats. In selecting a coating for hot structures, the engineer will also have to keep in mind two primary criteria: the ability of the coating to protect the component from degradation and the ability of the coating to retain, over the long term, its protective properties in the harsh environment of the engine.

For decades, improvements in gas-turbine efficiency, whether for land-based power generation, marine applications, or aircraft propulsion, have been primarily driven by significant increases in gas temperatures within the engine. Of particular note have been the dramatic increases in the gas-path operating temperature within the combustor and high-pressure turbine sections of the engine. For example, turbine inlet operating temperatures have increased by hundreds of degrees, and the specific thrust has nearly doubled during the five decades of the existence of gas-turbine engines.

These large increases in gas temperature have been enabled by the development of highly engineered cooling schemes for the substrate and high-performance materials technologies. Advances in cooling technology involve sophisticated convection, impingement, and film cooling designs for turbine hardware. In many cases, these developments were facilitated by parallel advances in materials processing technologies, such as precision casting, machining, and drilling operations. Cooling has arguably accounted for more than 75 percent of the increased temperature capability and is still the "first line of defense" in new turbine designs.

The introduction of new materials and processes has also heralded a series of advances in engine technology. For example, the transition from wrought to cast alloys provided a step improvement in high-temperature strength, a capability further enhanced by the introduction of vacuum investment casting. Within the range of cast superalloys, further incremental improvements occurred as directionally solidified and single-crystal alloys were developed to tolerate severe, cyclic thermal gradients. Future improvements will occur as intermetallic compounds, ceramics, composite materials, and refractory metal alloys are eventually incorporated in advanced engines, although significant technical barriers must still be overcome before these new families of materials are incorporated in critical engine components.

Because seemingly modest increases in upper temperature limits translate into huge benefits for engine performance, the large development costs associated with the introduction of new turbine blade alloys have historically been readily justified for even modest increases in temperature capability.[1] The cycle time associated with the introduction of a new alloy has resulted in an average improvement in temperature capability of approximately 5°C (9°F) per year.

As highlighted in this report, advances in superalloy materials have gone hand-in-hand with continual improvements in the coatings used to protect the components from environmental degradation mechanisms, particularly high-temperature oxidation and hot corrosion. The coatings used to date have not increased the inherent temperature capability of the substrate material, but they have protected superalloys from increasingly aggressive service environments and appreciably extended the service lives of critical, costly components. This coating/substrate combination has provided such improved component capability that, along with advanced cooling technology, turbine gas-path operating temperatures in localized regions exceed the melting point of the engine's structural materials without causing catastrophic failure.

Recently, ceramic thermal barrier coatings (TBCs) deposited onto turbine alloys have been shown to have two major benefits. First, for the same gas-path operating temperatures, the cooling air requirements can be greatly reduced. Second, for the same cooling air flow, the hot gas temperature can be raised significantly. In practice, the designer can choose the operating conditions between these two alternative ranges. Turbines can be operated at higher temperatures while maintaining an acceptable temperature range within the substrate

---

[1] Each generation of superalloy can cost $10 million or more. This rough cost estimate assumes that the new generation is based on an incremental materials and processing change from the previous generation, however. If a step change is required, such as the change from casting turbine blades with uniaxial grains to single crystals, the development costs would be much greater.

INTRODUCTION

material (i.e., TBCs are capable of providing at least 110°C [200°F] of increased capability). This latter benefit will allow TBCs to effectively fill the gap in materials capability for advanced engines between the current family of nickel-base superalloys (which are generally regarded as being at or close to their ultimate service temperature) and the advanced structural materials, such as intermetallics and ceramics.

As the gas-turbine industry becomes more dependent on coatings, the scatter (variation) in coating properties is becoming increasingly important. Figure 1-1 illustrates the scatter in performance of a given property of a coating, together with the property values and variation required to meet a set of advanced engine requirements. A wide bell-shaped distribution in a property such as coating adherence is tolerable—although less than optimal—when the coating is only to be used for extending the life of the base metal. This is particularly true if the failure mode is *graceful,* as is the case when a coating is protecting a superalloy that possesses inherent (but insufficient) environmental resistance.

If the coating must be relied on to last for a guaranteed time between overhauls, however, the performance of the system will be compromised if the coating cannot provide protection on a continued, reliable basis. If the coating system was further required to extend the temperature capability of the component system (as in the case of TBCs), scatter and unpredictability of the coating properties become critical. In either case, a coating with too great a variation in critical properties would not be implemented. For example, TBC life can be modeled as a broad statistical distribution, such as seen in figure 1-1. As a result, TBCs have generally not been used for critical applications such as for high-performance turbine airfoils. A current design criterion requires that a component not fail prior to the next scheduled inspection after local TBC spalling (Soechting, 1994). This limits the incremental thermal advantage of TBCs to approximately 40°C (70°F) over an uncoated part. This gain is nevertheless a significant benefit, approximating the gains typically achieved by a generation or two of alloy development. The full potential of TBCs can only be realized, however, if these coatings can be relied on to provide thermal protection of the substrate throughout the service life of the component. Thus, the coating must remain adherent and functional for the extended periods between coating refurbishment.

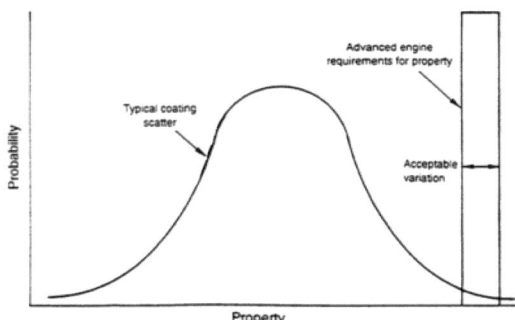

Figure 1-1 Typical ranges of a coating property compared with advanced engine requirements for a property.

The dependence on the reliability of coatings to improve performance and extend service life places an entirely new set of requirements on coating application processes and on coating system design. Coatings are becoming an integral part of the materials system needed to satisfy the application requirements and component design. *Critical objectives of coating technologists, therefore, are to reduce the scatter in the property distribution curve (i.e., to enhance reliability) and to move the minimum predicted value into the regime required by the advanced engine requirements.* This *theme* provides the basis for much of the discussion in this report and is a fundamental consideration of coating development and implementation on current and future high-temperature structural materials.

2

Application Needs and Trends

The operating requirements of gas-turbine engines dictate the future development of coatings. This chapter discusses two areas of that influence: application needs and substrate material trends. A cross section of a typical modern gas-turbine engine is shown in figure 2-1. As a general rule, components are only coated if system durability and reliability can be improved. For instance, within the last several years, high-temperature coatings have been extended to additional hot-section components in order to reduce surface degradation.

APPLICATION NEEDS

The need for coatings has evolved since they were first applied to high-pressure turbine airfoils in the early 1960s. Coatings were initially used to protect hot structure against degradation by high-temperature oxidation and hot corrosion (These processes are described in chapter 4). Coatings have become increasingly critical as advances in turbine technologies have made possible substrate and gas-path temperatures that approach or even exceed the melting point of the substrate in localized regions.

The duty cycles imposed on gas-turbine engines vary greatly depending on application. Gas-turbine engines are used in a variety of transportation, power-generation, and mechanical-drive applications, listed in figures 2-2 and 2-3. Operating life expectancy is principally determined by balancing the two major factors that contribute to wear: on/off cycle requirements and time-at-temperature. Greater demand on cyclic capability generally reduces the time a system can operate at high power (i.e., the interaction of thermomechanical fatigue and high-temperature creep of materials determine hot-section life expectancy). Other considerations, such as environmental factors (e.g., hot corrosion and high-temperature oxidation), also lower life expectancy.

While engine mission profiles vary with customer requirements, the typical hours of operation expected per engine start are depicted for each generic class of operation in table 2-1. The number of engine start/stop cycles and operating time at high temperature are the basic factors that determine when critical components are to be overhauled. The condition of hot-section coatings is an important indicator of the extent of distress suffered by an engine. Thus, the coatings are regularly inspected. The reliability and life expectancy of high-temperature coatings are major factors in maximizing the service life of a gas-turbine engine.

Aircraft Turbine Needs

Two government programs, the High-Speed Civil Transport program led by NASA and the Integrated High-Performance Turbine Engine Technology program led by DOD, are aimed at developing prototype turbines within the next ten years that will operate at firing temperatures at least 150°C (270°F) higher than the current generation of high-performance gas turbines. Figure 2-4 depicts the increase in firing temperatures over the past 15 years and the nominal goals for these programs. The committee believes that incremental improvements in the durability of substrate materials will not meet the goals of these two government programs. Success

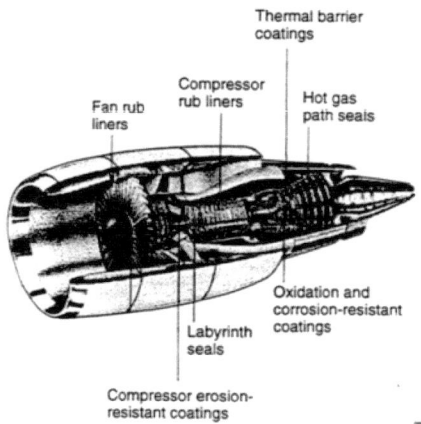

Figure 2-1 Cross section of a typical modern gas turbine engine.
Source: Hillery (1989). Copyright General Electric Company; used with permission of the General Electric Company.

APPLICATION NEEDS AND TRENDS

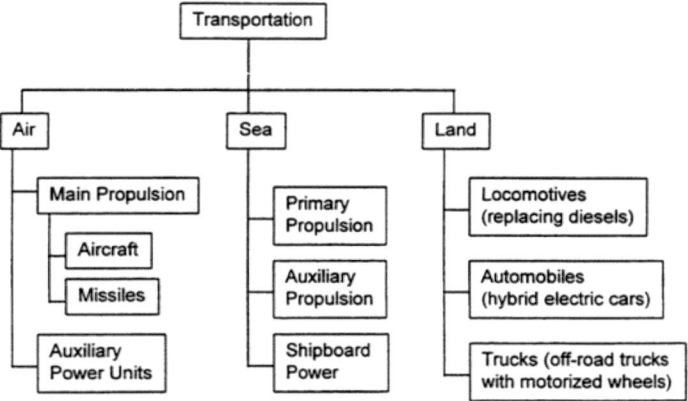

Figure 2-2 Overview of transportation gas-turbine use.

will require a much more rapid improvement in the operating temperature of materials, as depicted in figure 2-4. Incremental development of TBC technology, however, may allow the current generation of materials to achieve a portion of the desired increase in firing temperature.

In addition to a hot-section temperature increase, both of the advanced engine projects also seek to reduce emissions dramatically, particularly NO_X. These emission requirements affect the engineering design of the combustor and cannot be met without advances in combustor materials and coating technology. For example, release of NO_X and SO_X could potentially be controlled by the engineering of ingredients in the coating that act as catalysts for the conversion of these pollutants immediately after their formation.

Land-Based Turbine Needs

Industrial gas turbines for power generation are distinguished from aircraft gas turbines by many characteristics that include larger size, lower rotational speeds, longer life expectancy, longer time between overhauls, high time at full power, few on/off duty cycles, lower quality and lower heating-value fuels, wide variety of fixed locations (e.g., arctic to tropic, oceans to deserts, clean to polluted), and few weight penalties. The needs of large power-generation equipment essentially determine the available technology for land-based turbines (NRC, 1986).

The Advanced Turbine Systems (ATS) program, funded as part of the fossil energy research program of DOE, seeks to develop higher-efficiency land-based turbines that burn

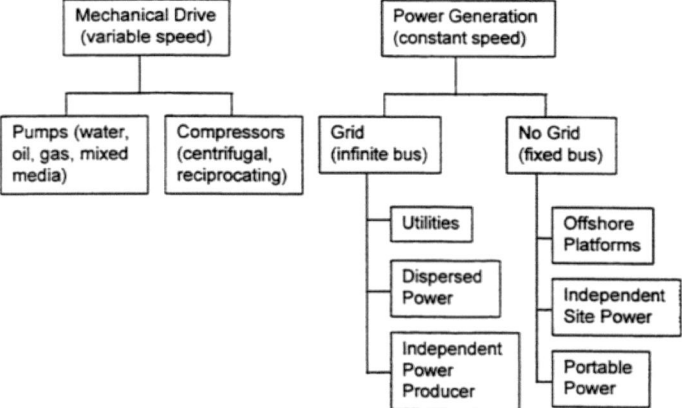

Figure 2-3 Overview of power-generation and mechanical-drive gas-turbine use.

APPLICATION NEEDS AND TRENDS

TABLE 2-1 Typical Duty Cycles for Various Gas-Turbine Engines

Duty Cycle	Typical Mission Time per Engine Start (hours)		
	Minimum	Average	Maximum
Airline/Transport	1	2	8
Military/Fighter[a]	1	2	4
Mechanical Drive	40	2,000	8,000
Base Power Generators	500	2,000	8,000
Peak Power Generators	2	4	12

[a] The mission profile for military aircraft has greater thermal transient and low-cycle fatigue because of periods of supersonic cruise and combat/avoidance maneuvers.

natural gas but that are also capable of burning coal-or biomass-derived fuels (NRC, 1995a). Firing temperatures at least 150°C (270°F) higher than the current generation of gas turbines and a combined cycle efficiency of 60 percent are desired, while maintaining the durability levels of current power-generation systems.

Fuel type is probably the most important variable affecting the selection of a coating system (NRC, 1995a). Pipeline-quality natural gas contains virtually no hydrogen sulfide and very low sulfur content, unlike unprocessed natural gas. Coal gasification produces a raw syngas consisting mainly of carbon monoxide and hydrogen, along with substantial quantities of hydrogen sulfide, ammonia, and hydrogen chloride, and a few parts per million of alkali metals (NRC, 1986). To meet environmental requirements, commercial gasification systems include clean-up systems that remove virtually all of the hydrogen sulfide, ammonia, and hydrogen chloride; in general, the alkali metals are also removed. However, it is possible that in the event of a degradation of the clean-up systems, some of these harmful materials may enter the turbine. The complex chemical and electrochemical reactions that occur at high temperatures and the susceptibility of these reactions to small changes in the coating and gaseous environment make coatings for hot-section components a critical enabling technology. Coating degradation by hot-corrosion mechanisms is the primary concern. One goal of the ATS program is to seek lower emission levels, similar to those discussed above for aeronautical turbines.

Marine Turbine Needs

On the ocean, gas turbines provide primary drive, auxiliary drive, and shipboard power. A new type of marine vehicle,

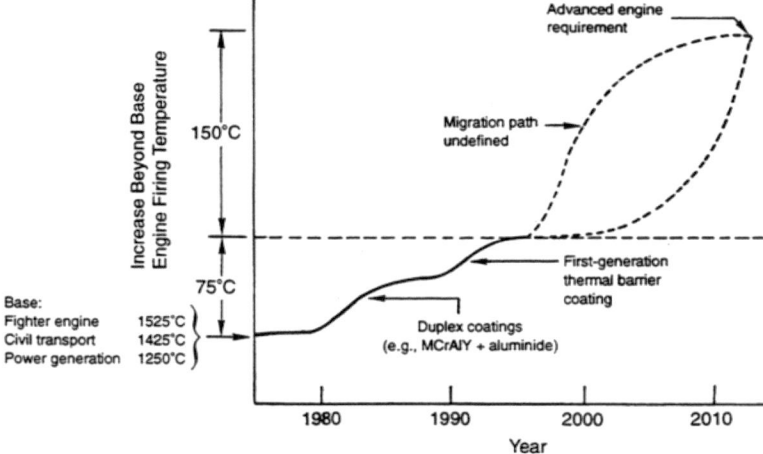

Figure 2-4 Approximate increases in firing temperature capabilities from 1980 to 2010.

APPLICATION NEEDS AND TRENDS

the all-electric ship, uses a gas turbine located at mid-ship to power the motors that drive the aft propellers. Because of their low operating cost, gas turbines are expected to become more common on ships.

Although high-quality liquid fuels are used, marine gas turbine engines operate in an environment heavily laden with salt (sodium chloride) and are often close to other ships that burn low-quality fuels with high-sulfur concentrations, such as Bunker C. Contaminants from these exhaust gases are drawn into the ship's intake air. Thus, hot corrosion is the primary nemesis of gas turbines on ships. The coatings primarily developed for marine environments are aimed at increasing the resistance of hot-section components to hot corrosion.

CURRENT COMPONENT DESIGN AND BASE MATERIALS

The operating conditions that a component is expected to encounter drives the component functional design and the resulting choice of material. The materials that have been and continue to be used are nickel-base and cobalt-base superalloys. A partial list of these alloys is given in table 2-2 (Gabb and Dreshfield, 1986; Backman and Williams, 1992; Stringer and Viswanathan, 1993). These chemically complex alloys are highly developed materials with a combination of high temperature strength, inherent environmental resistance, toughness, and fatigue resistance that are not currently available in other alloy system (Sims, 1986). Numerous compositions and processing routes have been developed that can be tailored to specific application needs.

Nickel-base superalloys have had more extensive development than cobalt alloys. Nickel-base superalloys enjoy high strengths at elevated temperatures through the precipitation of a coherent phase: gamma-prime. In addition, these alloys are strengthened by the more common mechanisms of incoherent phases and solid solutions. The result is a family of chemically complex alloys that are capable of operating in aggressive environments at temperatures greater than 90 percent of their incipient melting temperature. These features have made nickel-base superalloys the favored materials for use in rotating turbine components, such as blades. These alloys are also widely used in static structure, such as case, vane, and combustor applications.

The dramatic progress in investment-casting technology ranks as the single most important process development for nickel-base superalloys. Progress in this technology has enabled engineers to cast directionally solidified and single-crystal airfoils that include complex, cast-in, internal air-cooling passages. By using sophisticated cooling techniques with single-crystal alloys, researchers have attained operating temperatures 80°C (144°F) higher than those of conventional, equiaxed castings.

Cobalt-base superalloys are not as developed as nickelbase superalloys because coherent phase strengthening has not been demonstrated for these materials. Although cobaltbase superalloys have less strength at comparable temperatures, they tend to have higher melting points than comparable, conventionally cast nickel-base superalloys and can therefore function at somewhat higher temperatures. Cobalt-base superalloys are more easily repaired by welding than are the advanced nickel-base superalloys and also have better corrosion resistance because of higher chromium levels. For these reasons, the cobalt-base superalloys are used for higher-temperature applications in which high strength is not the primary issue (i.e., vanes and combustor liners). These alloys also prove valuable in certain-but not all-hot-corrosion conditions. However, the advanced nickel-base alloys developed for single-crystal technology have significantly higher melting temperatures than the conventional nickel-base alloys. As a result, single-crystal nickel-base superalloy vanes have replaced polycrystalline cobalt vanes in some applications.

Design engineers seek to mitigate the impact of operating conditions on component integrity by two approaches, both of which have become absolutely essential to the design of hot-section components. The first approach relies directly on coating technology to extend the service life of both nickelbase and cobalt-base superalloys, as discussed in chapter 3.

In the second approach, coatings contribute indirectly, but just as significantly, to extending service life. This method relies on component cooling to decrease the steady-state component temperatures; coatings play a key role in such cooling. For aeronautical turbines, noncombusted air diverted from the compressor is used to cool critical components, such as airfoils (Sims, 1991). Cooling through the use of steam or other means is also possible when engine weight is not a concern. Several methods working in tandem make the most efficient use of the cooling air. These include convection cooling and impingement cooling on the back sides (cold side) of hot surfaces and film cooling on the hot sides that face the combustion gases (figure 2-5). In film cooling, air flowing out through small holes in the component wall provides a boundary layer of air cooler than the combustion gas, thus cooling the component. These film-cooling holes may become blocked during the application of a coating, however. Loss of cooling air may lead to local melting and serious degradation of the component.

ADVANCED MATERIALS FOR ENGINE COMPONENTS

Nickel-base and cobalt-base superalloys have served admirably in gas turbines for over 50 years, and the attractive properties of this class of materials guarantee that superalloys will continue to serve in turbine engines well into the future (Sims, 1986). However, current superalloys operate at 90 percent

TABLE 2-2 Nickel- and Cobalt-Base Alloys Used in Land-Based and Aircraft Gas Turbines

Combustor Liners and Transition Pieces: Sheet Metal Applications

Alloy	Ni	Co	Cr	Al	Mo	W	Ti	Fe	Cb	C	Si	Mn	Zr	Ta	Other	Type	$T_{SOLIDUS}$ (°C)
Hastalloy X	Bal.	—	22	—	9	—	—	20	—	0.15	—	—	—	—	—	sheet	1260
Haynes 188	22	Bal.	22	—	—	14	—	3 max.	—	—	—	—	—	—	—	sheet	1300
TD-Nickel	Bal.	—	—	—	—	—	—	—	—	—	—	—	—	—	2.0 ThO$_2$	sheet	1453
Inconel 617	Bal.	12.5	22	1	9	—	—	—	—	0.07	0.5	0.5	—	—	—	sheet	1330
RA 333	Bal.	3	25	—	3	3	—	18	—	0.05	1.25	1.5	—	—	—	sheet	*

Stator Vanes and Nozzles

Alloy	Ni	Co	Cr	Al	Mo	W	Ti	Fe	Cb	C	Si	Mn	Zr	Ta	Other	Type	$T_{SOLIDUS}$ (°C)
X-40	10.5	Bal.	25.5	—	—	7.5	—	—	—	0.5	—	—	—	—	—	CC	1341
Mar-M 509	10	Bal.	23.5	—	—	7	0.2	—	—	0.6	—	—	0.45	3.5	—	CC	1288
F-75	—	Bal.	27.5	—	5.5	—	—	—	—	0.22	—	—	—	—	—	CC	*
GTD 222	Bal.	19	22.5	1.2	—	2	2.3	—	0.8	0.1	—	—	—	1	0.12N	CC	1260
IN713C/HC	Bal.	—	13.5	6	4.5	—	0.85	—	2.3	0.12	—	—	0.1	—	—	CC	1288
IN713C/LC	Bal.	—	13.5	6	4.5	—	0.85	—	2.3	0.06	—	—	0.1	—	—	CC	*
IN738C/HC	Bal.	8.5	16	3.4	1.7	2.6	3.45	—	0.9	0.17	0.3	0.2	0.1	1.75	0.012B	CC	1232
IN738C/LC	Bal.	8.5	16	3.45	1.7	2.6	3.6	—	0.9	0.06	0.3	0.2	0.1	1.7	0.012B	CC	1232

Turbine Blades

Alloy	Ni	Co	Cr	Al	Mo	W	Ti	Fe	Cb	C	Si	Mn	Zr	Ta	Other	Type	$T_{SOLIDUS}$ (°C)
B1900	Bal.	10	8	6	6	—	1	—	—	0.1	0.2	0.2	0.07	4.25	0.015B	CC	1274
IN713C	Bal.	—	13.5	6	4.5	—	0.85	—	2.3	0.12	—	—	0.1	—	—	CC	1260
IN792	Bal.	9	12.5	3.3	1.9	4.2	4	—	—	0.08	—	—	0.03	4.2	0.015B	CC	*
IN738	Bal.	8.5	16	3.45	1.7	2.6	3.45	—	0.9	0.11	0.3	0.2	0.1	1.75	0.012B	CC	1232
RENE 77	Bal.	15	14.6	4.3	4.2	—	3.35	—	—	0.07	—	—	—	—	0.015B	CC	1093
MAR-M 200Hf	Bal.	10	9	5	—	12.5	2	—	1	0.15	—	—	—	—	2.0Hf	DS	1316
RENE N4+	Bal.	7.5	9.7	4.2	1.5	6	3.45	—	0.5	0.06	—	—	—	4.8	0.15Hf 0.004B	SC	*
RENE N5	Bal.	8	7	6.2	2	5	—	—	—	—	—	—	—	5	3Re	SC	*
CMSX-2	Bal.	4.7	7.9	5.5	0.6	8	1	—	—	—	—	—	—	—	—	SC	*
CMSX-4	Bal.	9.5	6.2	5.5	0.6	6.5	1	—	6	—	—	—	—	—	2.9 Re 0.1Hf	SC	*
PWA1480	Bal.	5	10	5	—	4	1.5	—	6.5	—	—	—	—	12	—	SC	*
PWA1484	Bal.	10	5	5.6	2	6	—	—	—	—	—	—	—	8.7	3.0 Re 0.1Hf	SC	*

*Data Unavailable

NOTES: Compositions are in weight percent.
Alloy "type" refers to the fabrication process: CC = conventionally cast; DS = directionally solidified; and SC = single crystal.

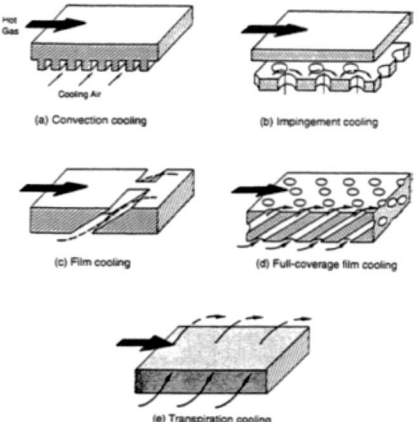

Figure 2-5 Turbine blade cooling methods. Source: Glassman (1975).

of their incipient melting temperatures in some applications. This is both an unprecedented achievement and a graphic illustration of the inherent limitation of these materials. The requirement for more efficient engines in the future will also require even higher operating temperatures and, therefore, higher component temperatures. Thus, the targeted temperatures will inevitably exceed the capabilities of current superalloys.

Variants of current superalloys, such as eutectics and superalloy matrix composites, can fill a need for increased creep resistance and higher specific strength. These materials can also help reduce component weight and, when used as turbine blades, lower the loads on turbine disks. However, their potential for withstanding higher temperatures is still limited by their melting points. Thus, designers are investigating alternative materials to fill the higher-temperature roles. These materials generally fall into three classes: ceramics, intermetallics, and refractory materials (Stoloff and Sims, 1986; Meetham, 1988; Backman and Williams, 1992; Stringer and Viswanathan, 1993). The remainder of this chapter discusses the environmental degradation issues that are faced for each class of these materials. Coatings for these materials are presented in chapter 3.

Ceramics

Ceramics are generally thought to hold considerable promise because they possess many of the properties desired for higher-temperature replacements of superalloys: very high melting points, high-temperature strength, low density, and some increased resistance to aggressive environments (table 2-3). The primary shortcoming of monolithic ceramics is their lack of acceptable low-temperature ductility and toughness. The toughness issue is being addressed by the development of ceramic composite systems.

Ceramics currently under development may serve as combustors in both aircraft and land-based turbine engines in the Advanced Turbine Systems and the Integrated High-Performance Turbine Engine Technology programs. Designers are also exploring the use of ceramics for turbine blades, an application that could become the biggest economic payoff of these materials. Silicon carbide (SiC) and silicon nitride (Si_3N_4) rank as the prime candidates for making ceramic components because of their high-temperature strength and relative resistance to thermal shock.

Although the development of the requisite mechanical properties of ceramic components continues, it is becoming increasingly clear that these materials must be coated. Both silicon carbide and silicon nitride oxidize to form silicon oxide surface scales, which may limit their high-temperature use unless adequate protective coatings can be applied. Under the operating conditions envisioned for these materials, surface recession caused by active oxidation (SiO formation instead of SiO_2) is a significant problem (Jacobson, 1992). Even limited exposure to sodium-containing environments greatly accelerates active oxidation (Jacobson et al., 1990). Coatings for a ceramic component will almost surely have to be ceramic as well, which poses the problem of having to precisely match coefficients of thermal expansion since ceramics do not possess the ductility to accommodate even slight mismatches in thermal expansion. In summary, coatings are needed to act as both thermal barriers to reduce the base-material temperature (and thereby reduce the oxidation rate) and as chemical barriers to slow oxidation.

Oxide-base ceramics are not as attractive for high-temperature structural applications since their strength and toughness are not as high as silicon-base materials (Stoloff and Sims, 1986). Coatings are less important for these materials, however, since oxidation is clearly not a concern, although some (e.g., zirconia and alumina) are susceptible to hot corrosion.

Intermetallics

Intermetallic compounds, especially those involving aluminum, have been heavily researched over the last several years. Compounds of nickel (Ni), titanium (Ti), cobalt (Co), and iron (Fe) with aluminum (Al) have all shown promise, but most studies have focused on nickel aluminides and titanium aluminides. These compounds have higher melting temperatures and lower densities than the superalloys (table 2-4). Thus, they have the possibility of achieving both higher temperatures and higher specific strengths than superalloys. Yet even these alloys will require coatings to excel at high temperatures.

TABLE 2-3 Properties of Ceramic Materials that are Candidates for Hot-Section Use

Material	$T_{MELTING}$ (°C)	CTE (10^{-6} °C^{-1}) to ≈1000°C	Young's Modulus @ 21°C (GPa)	Density (g/cc)	Fracture Toughness[a] @ 21°C (MPa √m)	Flexural Strength[a] @ 21°C (MPa)	Flexural Strength @ 1200°C (MPa)
SiC	[b]	4.0[c]	410[d]	3.2[c]	3	400	276–690[c]
Si$_3$N$_4$	[b]	3.0[c]	310[d]	3.2[c]	4.0–5.5	500–800	550–830[c]
Composite Si$_3$N$_4$ 30 vol % SiC whiskers	—	—	—	—	6.5–7.5	700–950	—
Al$_2$O$_3$	2045[e]	9.0[f]	360[d]	3.97	2.5–4.0	300–400	—
Composite Al$_2$O$_3$ 20 vol% SiC whiskers	—	—	—	—	7.5–9.0	650–800	—
ZrO$_2$	≈2700[f]	10.0[f]	200[d]	≈5.73	—	—	—
fully stabilized	—	—	—	—	2.3	100–300	—
partially stabilized	—	—	—	—	15–18	>600	—
TZP	—	—	—	—	5–16	>1000	—
B 1900 Superalloy[d]	1330	15.8	214	8.0	≈100[g]	970	270 (UTS @ 1100°C)

[a] Tiegs et al. (1992).
[b] Silicon-base ceramics tend to decompose and sublimate at high temperature. Therefore, there is no definable melting point for atmospheric pressures. An "optimistic maximum use temperature," based on oxidation behavior, is 1650°C (Fox, 1992).
[c] Larsen et al. (1985).
[d] CRC (1985).
[e] International Nickel Company (1977).
[f] Kingery et al. (1976).
[g] Actual values for B 1900 were not available. This represents typical superalloy toughness values for comparison with ceramics.

NOTE: Ceramic properties are very sensitive to processing; therefore, the values given are representative of a range of values for each property.

Researchers have also begun investigating intermetallics as matrix materials for composites, adding to the specific strength capabilities of the intermetallics. Furthermore, alloys with higher aluminum content (e.g., NiAl, TiAl, and TiAl$_3$) have good inherent resistance to oxidation. The oxidation resistance of NiAl is particularly excellent; NiAl is the main component in oxidation-resistant aluminide coatings for nickel-base superalloys.

Although researchers have considered intermetallics for structural applications for over 20 years, most of the studies on these materials have been conducted during the last decade (Dimiduk et al., 1992). Therefore, intermetallics remain relatively immature materials that have a significant number of-technical barriers to be overcome before they could be used as hot-section materials. Obstacles include low creep strength at high temperature and poor ductility at low temperatures(Dimiduk et al., 1992). Despite these drawbacks, a number of research programs have improved the properties of these materials, and intermetallics now offer the promise of less-demanding, lower-temperature applications. Indeed, studies appear to have shifted from higher-temperature (roughly between superalloy and ceramic capabilities) to lower-temperature applications where these materials have a chance of replacing the denser superalloys. As was the case for superalloys, intermetallics may only attain the required mechanical properties at the expense of some of their environmental durability. Thus, intermetallics developed for hot-section uses will require coatings to mitigate a variety o f environmentally induced problems.

Titanium aluminides that have low aluminum content (i.e., less than 40 atomic percentage) tend to grow brittle during exposure to high temperatures in air. The high-aluminum-content titanium aluminides (TiAl) apparently do not have as severe an embrittlement problem and have a higher resistance to oxidation than the low-aluminum titanium aluminides (Ti$_3$Al; McKee and Huang, 1990). Unfortunately, the relatively good oxidation resistance of TiAl does not extend to the higher temperatures required in the hot section. Thus, TiAl will probably require a coating to perform adequately in the hot section. The oxidation-resistant TiAl$_3$ is brittle at low temperatures and is not currently a promising structural material (Smile et al., 1990).

Applications for NiAl are still somewhat tentative since alloys having the requisite mechanical properties for hot-section use are not currently available. Near-stoichiometric NiAl has excellent resistance to oxidation and therefore should not need an oxidation-resistant coating. However, the poor strength of pure NiAl at high temperatures suggests that the

TABLE 2-4 Selected Properties of High-Temperature-Capable Intermetallics Compared to a Conventionally Cast and a Single-Crystal Superalloy

Material	$T_{SOLIDUS}$ (°C)[a]	Young's Modulus (GPa)	Density (g/cc)	Fracture Toughness @ 21 °C (MPa`m)
Ti_3Al[b]	1600	145	4.3	25
TiAl[b]	1460	176	3.8	25
NiAl[a]	1640	300	5.9	—
NiAl-eutectic[b]	—	193	5.9	12-15
$MoSi_2$[b]	1870-2030	379	6.5	4-5
FeAl[a]	1250	260	5.6	—
CoAl[a]	1650	300	6.1	—
B 1900[c]	1274	214	8.2	»100
Single-crystal superalloy[a]	—	125	8.3	35

[a] Stoloff and Sims (1986).
[b] Dimiduk et al. (1992).
[c] International Nickel Company (1977).

Other intermetallics under investigation for high-temperature use include $MoSi_2$, FeAl, and CoAl. It is expected that the same concerns about hot-section oxidation resistance and environmentally induced problems will apply to FeAl and CoAl. Although its high-temperature oxidation resistance is excellent (Vasudevan and Petrovic, 1992), $MoSi_2$ has the added environmental-resistance problem of pesting.[1]

material will need to be alloyed for use in a hot section. Alloying will most likely cause some loss of oxidation resistance. Coatings may thus be required to provide requisite oxidation resistance.

Refractory Metals

Scientists have gathered extensive data over the last few decades on the properties and environmental resistance of the refractory materials (Stoloff and Sims, 1986). The continued interest in these materials stems from the attractive combination of high melting point, similar to (and in some cases, higher than) ceramic materials, with some ductility at low temperatures (table 2-5). Most of the research has been conducted on niobium, molybdenum, tantalum, and tungsten because of their relatively low cost compared with other refractory metals. Niobium is the most attractive of the refractories since it has a density slightly lower than nickel while still having a melting temperature approximately 1000°C (1800°F) higher than nickel. In fact, the niobiumbase alloy C-103 has been successfully used in unmanned aircraft (Stoloff and Sims, 1986). Although a diffusion coating of (molybdenum/tungsten)(silicon/germanium)$_2$ has reportedly protected niobium (Mueller et al., 1991), the extremely high oxidation rates and tendencies toward embrittlement of the refractory alloys would result in extremely rapid degradation if the coating failed. Consequently, these materials are not currently used in long-lived, hot-section components.

Recent work on refractory alloys has also focused on developing resistance to environmental attack. These metals have little inherent resistance to environmental attack, which could result in a coating flaw leading to the degradation and subsequent failure of a component. This work has included additions of aluminum and titanium, both of which tend to slow oxidation, and additions of sufficient aluminum to form refractory metal aluminides (Stoloff and Sims, 1986). These efforts have not produced alloys useful as hot-section components in air-breathing engines, however. If these efforts ultimately prove successful, a coating will probably still be necessary to provide the long-term environmental resistance required for all gas-turbine engines.

---

[1] Pesting is a thermally activated process observed in several intermetallic compounds that results in catastrophic degradation of the material. At an intermediate temperature range that is system specific, accelerated oxidation of the intermetallic causes large volume increases. When oxide growth occurs in cracks, porosity, and grain boundaries, the large volume increase results in fracture and eventually catastrophic failure of the material (Doychak, 1994).

TABLE 2-5 Properties of Refractory Metals and an Alloy[a] Compared to B 1900 Nickel-Base Superalloy

Material	$T_{SOLIDUS}$ (°C)	CTE x 10^6 (°C)	Density (g/cc)	Young's Modulus @ 25°C (GPa)	UTS (MPa)
Niobium	2468	8.3	8.6	97	—
Alloy B88 (Nb-28W-2Hf)	—	—	»11.6	—	»462 @ 1315°C
Tantalum	2996	6.7	16.6	185	—
Molybdenum	2610	5.8	10.2	325	—
Tungsten	3410	4.6	19.3	400	—
B 1900[b]	1274	15.8	8.0	214	270 @ 1100°C

[a] Stoloff and Sims (1986).
[b] International Nickel Company (1977).

3

Materials and Processes

As described in chapter 2, the primary purposes of high-temperature structural coatings are to enable high temperature components to operate at even higher temperatures, to improve component durability, and to allow use of a broader variety of fuels in land-based and marine-based engines. Although high-temperature coatings protect the substrate, the demarcation between coating and substrate (either metal or nonmetal) is becoming increasingly blurred. The demanding requirements of high-temperature service in both isothermal and cyclic modes have recast the way researchers think about coated structures. These structures can be considered part of a continuum; at the limit the coating will be a progressive modification of the substrate and therefore must be concurrently designed with the substrate. Table 3-1 summarizes the relationship between coating functions and coating characteristics.

There are essentially two types of high-temperature coatings. The first type is a diffusion (or conversion) coating in which the deposited mass is diffused and/or reacted with the substrate to form a somewhat continuous gradation in composition. The second type is an overlay coating in which material is deposited at the surface of the substrate. This chapter discusses current and potential coatings for superalloys, ceramics, refractory metals, intermetallic materials, and metal-matrix composites. The processes used to apply these coatings to substrate materials are also reviewed.

TABLE 3-1 Coating Functions and Coating Materials Characteristics

Functions	Materials Characteristics
Reduction in surface temperature	Low thermal conduction Low radiative heat transfer High emittance
Reduction in rate of oxidation	Thermodynamically stable oxide formers with slow growth rates
Reduction in rate of hot corrosion	Chemically stable and impervious oxide scale
Resistance to particulate erosion	Hard, dense material
Increased abradability (sacrificial wear)	Rub tolerance via plastic deformation (densification)
	Energy transformation via fracture (material loss)
Increased abrasiveness	Inclusion of hard particles to induce cutting of seal

COATINGS FOR HIGH-TEMPERATURE STRUCTURES

Coatings must maintain their performance on a continued and reliable basis, or the performance of the turbine system could be compromised. Advanced coatings that show excessive variability and unpredictability at the required engine operating conditions will not be employed, regardless of their potential benefits to the system. For example, thermal barrier coatings (TBCs) are currently used to improve the performance of several engines, including GE's CF6-80C2 and CFM56-5a and Pratt & Whitney's PW2000 and PW4000 series engines (Bose and DeMasi-Marcin, 1995). TBCs are currently relied on for thermal insulation to improve component durability at elevated temperatures (i.e., 40°C [72°F] higher than uncoated parts) but still below the incipient melting temperature of the substrate material. In addition, TBC patches are used to minimize thermomechanical fatigue cracking by reducing substrate temperatures at component hot spots. However, TBCs will not be widely accepted for use at temperatures that will result in the rapid degradation of substrate materials in order to provide the performance benefits of which they are capable until they can be produced with predictable, reliable properties, as discussed in chapter 1 (see figure 1-1).

The majority of hot-section components are currently made from superalloys. To achieve the aggressive performance goals of the advanced engine programs (e.g., High-Speed Civil Transport, Integrated High-Performance Turbine Engine Technology, and Advanced Turbine Systems programs) discussed in chapter 2, material systems with higher inherent temperature capabilities and better oxidation and corrosion resistance are required. These systems may require the use of advanced substrates, such as intermetallic compounds, ceramics, or refractory metals. TBCs deposited on current alloys may allow components to operate at higher temperatures.

TABLE 3-2 Types of Coatings Used in Hot-Section Components

Component	Coating Type		
	Aluminide	Chromide	MCrAlY
Blades			
Gas path	H L	H	M
Internal	M		
Tip		L	
Attachment	L		
Vanes			
Gas path	H L	H	M
Internal	M		
Shroud	H	H	M
Combustor			
Liner		M	M
Transition		M	M

NOTE: H= high use
M = medium use
L = low use

Coatings for Superalloys

In the past, coatings for superalloys formed a metallic-aluminide layer on the gas-path surfaces of high-pressure turbine airfoils. As alloys with improved temperature and cyclic capability were synthesized, suitable coatings were developed in parallel. These more advanced coatings were often tailored for expected environmental conditions.

Metallic coatings (i.e., aluminide, chromide, and MCrAlY) protect superalloys against aggressive environmental factors. Ceramic oxide coatings insulate the substrate from the maximum gas-path temperatures. Table 3-2 provides an overview of the types of coatings currently used for hot-section structures. Basic service use and application process information for each main coating type is summarized in table 3-3. In figure 3-1, Novak (1994) schematically depicts the tradeoffs in selecting coating compositions for an environment that requires resistance to both oxidation and corrosion.

In most applications, TBCs contain yttria-stabilized zirconia. TBCs are applied to an oxidation-resistant bondcoat, typically a MCrAlY designed for high-temperature oxidation resistance or an aluminide modified by a platinum addition (Smith and Boone, 1990). The platinum in an aluminide diffusion coating has been shown to improve the protective properties of the alumina scale by (1) increasing the purity of the alumina scale by prohibiting the diffusion of the refractory metals present in the substrate (e.g., tungsten and molybdenum) to the oxide scale; (2) increasing the diffusivity of aluminum and promoting alumina formation; (3) improving the selective oxidation of aluminum; and (4) decreasing the activity coefficient of aluminum in the coating. The bondcoat acts to retard a principal cause of TBC failure: oxidation of the interface between the coating and the substrate. However, alloys currently under development have shown improved oxidation resistance and may not require a bondcoat (Miller and Brindley, 1992; Ulion and Anderson, 1993). Furthermore, because of its additional weight, the bondcoat can be detrimental to the creep life of the component, particularly in the case of rotating parts. Research indicates that adherent, durable, plasma-sprayed zirconia-yttria TBCs can be deposited onto smooth substrates (without a metallic bondcoat) if an initial layer of the TBC is applied by low-pressure plasma spraying (Miller and Brindley, 1992). Current familiarity with TBCs will lead to improved design criteria and manufacturing experience that will allow continual improvement and additional benefits.

Coatings for Ceramics

Current approaches for coating the silicon-based ceramic materials include the use of mullite coatings that have an excellent expansion match with silicon carbide (Van Roode et al., 1992; Lee et al., 1994). While mullite contains silica, the aluminum in mullite promotes formation of a solid rather than a liquid oxidation product. This cuts the weight loss in half for

MATERIALS AND PROCESSES

TABLE 3-3 Generic Information on Coating Types Used in Superalloy Hot-Section Components

Information	Coating Type			
	Aluminide	Chromide	MCrAlY	Ceramic Oxide
Primary Protection Function	Oxidation	Hot corrosion	Oxidation	Thermal barrier
Thickness (μm)				
Nominal	65	25	125	250
Range	25-100	25-50	125-500	>125
Service Temperature				
°C	815-1150	600-925	815-1150	980-1200
°F	1500-2100	1100-1700	1500-2100	1800-2200
Application Process	*Vapor phase reaction*	*Vapor phase reaction*	Thermal spray	Thermal spray
			air/vacuum plasma combustion	air/vacuum plasma
	pack cementation above the pack chemical vapor deposition slurry	pack cementation above the pack	*Vapor deposition* electron beam sputtering *Laser deposition* powder cladding *Reaction sintering*[a] controlled composition	*Vapor deposition* electron beam sputtering
Relative Cost	1x	1x	2x-4x	3x-5x
Commercial Availability	Widely available	Widely available	Widely available	Selectively available

[a] Hsu et al. (1979).

silicon carbide materials under active oxidation conditions Multilayer coatings that incorporate mullite are currently under investigation (K. Lee, personal communication, 1996)

Coatings for Refractory Metals

Efforts to develop coatings for refractory metals have been hampered by their high reactivity. The aluminides (Stoloff and Sims, 1986) and silicides (Mueller et al., 1991) have produced coatings that reduce environmental-induced degradation. A new (molybdenum/tungsten)(silicon/germanium)2 diffusion coating on niobium has shown excellent cyclic oxidation resistance (Mueller et al., 1991) but has yet to be commercially developed. Use of even excellent coatings is unlikely, however, since flaws in the coating could severely curtail component life.

Coatings for Intermetallic Materials

In coating both the low-and high-aluminum-content titanium aluminides (i.e., Ti_3Al and TiAl), researchers may have

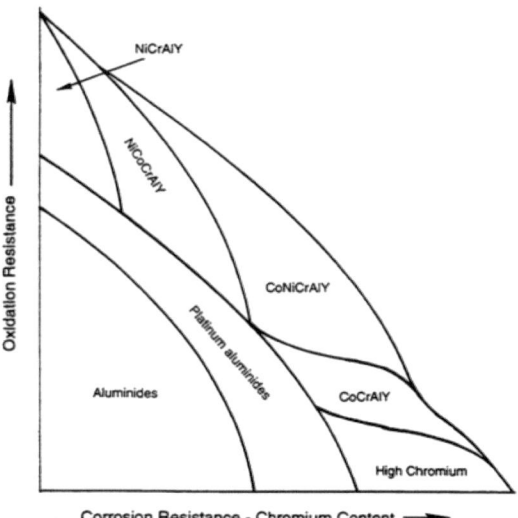

Figure 3-1 Coating compositions as related to oxidation and corrosion resistance. Source: Novak (1994).

to contend with coating/component interdiffusion and the potential for formation of brittle intermetallic compounds at the coating/substrate interface (Brindley et al., 1992; McKee, 1993). The high reactivity of the titanium-base alloys may require reaction barriers between the coating and the substrate to reduce the possibility of degradation (McKee, 1993). Flaws in either the coating or the reaction barrier could severely curtail component life.

The mechanical properties of nickel aluminide (NiAl) substrates are sensitive to compositional changes. Thus, the potential for diffusion between the coating and the component must be considered in the use of this alloy in the coated condition.

COATING PROCESSES

In general, the development of manufacturing processes for high-temperature coatings has paralleled the evolution of gas-turbine materials and component design. Early turbine blades for Allison and Curtiss Wright engines in 1952 were given a protective coating by simply dipping them into molten aluminum (Goward and Cannon, 1988). The first practical use of a diffusion coating for turbine airfoils was in 1957 (Goward and Cannon, 1988).

As interest surged during the 1970s, processes were developed for the application of overlay coatings that offered greater compositional flexibility than was possible with diffusion coatings. Most overlay coatings for gas-turbine applications are currently applied using electron-beam physical vapor deposition (EB-PVD), which is an atomistic deposition method, and low-pressure plasma spraying, which is a particulate deposition method. The benefits and limitations of the atomistic and particulate deposition methods are summarized in table 3-4.

The features, applications, advantages, and disadvantages of the coating methods most relevant to high-temperature structural materials are summarized in the remainder of this section. Appendix C cites a number of nondestructive evaluation methods that could be used for process control monitoring. Appendix E reviews examples of industrial manufacturing technologies for selected processes.

TABLE 3-4 Summary of the Benefits and Limitations of the Atomistic and Particulate Deposition Methods

Features	Processing Evaporation	Sputtering Deposition	CVD	Electrodeposition	Thermal Spraying
Mechanism to produce depositing species	Thermal energy	Momentum transfer	Chemical reaction	Solution	Flames or plasmas
Deposition rate	Moderate (up to 750,000 Å/min.)	Low	Moderate	Low to high	Very high
Deposition species	Atoms	Atoms/ions	Atms/ions	Ions	Droplets
Complex shapes	Poor line of sight	Good but nonuniform	Good	Good	Poor resolution
Deposits in small, blind holes	Poor	Poor	Limited	Limited	Very limited
Metal/alloy deposition	Yes	Yes	Yes	Yes	Yes
Refractory compounds and ceramics	Yes	Yes	Yes	Limited	Yes
Energy of deposit-species	Low	Can be high	Can be high	Can be high	Can be high
Growth interface perturbation	Not normally	Yes	Yes	No	No
Substrate heating	Yes, normally	Not generally	Yes	No	Not normally

Physical Vapor Deposition

The physical vapor deposition process is an atomistic deposition method that involves the vaporization and subsequent deposition of coating species. It has the advantage of being able to deposit coatings of metal, alloys, and ceramics on most materials and a wide range of shapes. Because application requires clear line-of-sight, complete coating coverage is achieved by manipulating the part during the coating cycle with a complex mechanical system.

Electron-beam guns for EB-PVD are favored for supplying the energy necessary for evaporation because they can achieve higher energy densities than other methods of heating. EB-PVD can successfully deposit mixed oxide coatings that are currently of greatest benefit to high-temperature structural materials (e.g., yttria-stabilized zirconia TBCs). In this case, the process parameters are adjusted so that the deposit has a columnar structure that is perpendicular to the interface. This morphology maximizes resistance to strains that stem from differences in thermal expansion coefficients. Because TBCs are porous, they allow gases and fused condensates to penetrate and, therefore, are deposited onto an underlying bondcoat resistant to oxidation and hot corrosion (usually MCrAlY).

Most models for EB-PVD processes are primarily empirical. EB-PVD equipment is computer controlled, and regulation of the process can often be devised by trial and error. Plasma and ion control of the vapor allows formation of the desired coatings. The committee was unable to learn much of the processing details of EB-PVD because this information is largely proprietary. Basic scientific understanding of vapor formation and deposition is known for simple systems but not for the more complex, multicomponent systems that generate superalloy coatings. Particularly troublesome are components with constituents that have vastly different vapor pressures. Deposition on parts with widely varying curvatures, such as airfoils, also presents difficulties in processing. The committee is not aware of any mathematical models for these situations that are available in the public domain. Additional information about this process is summarized in appendix E.

Sputtering

The primary advantage of sputtering is its ability to deposit a wide variety of materials (e.g., alloys, oxide solutions, and intermetallics). These compositions can be derived from many types of targets. A reactive gas can be introduced with the heavy inert gas, so that reactive sputtering occurs. Often the deposit has a columnar microstructure, with elongated grains normal to the interface. Such a structure would be ideal for oxide TBCs. Sputtering is not yet a production coating method for turbine hardware because current equipment deposits the coating too slowly. Appendix E provides additional information about this process.

Thermal Spray Processes

Thermal spray processes are particulate deposition methodologies that involve the deposition of molten droplets of material on a substrate. The coating typically begins as a powder. It is then injected into a hot carrier gas and sprayed on the target substrate with a gun. This process can quickly deposit a wide variety of coatings. A shortcoming of traditional thermal spray technologies has been a limitation on the thinness and smoothness of the deposit.

A wide variety of thermal spray techniques, as discussed in appendix E, can be employed to deposit metallic corrosion and oxidation-resistant overlay coatings, bondcoats, and ceramic TBCs. Plasma spraying is the most widely used thermal spray technique for gas-turbine component overlay coating.

The microstructure of a plasma-sprayed coating depends on the starting material and its particle size distribution, as well as processing parameters. New and improved powder processing methods can produce starting materials with predictable and controllable compositions and well-delineated particle size. Key plasma-spray process parameters include plasma power, plasma gas composition, pressures and flow rates, powder injection details and carrier flow, and torch/substrate distance. These parameters may be linked in complex ways, making process control difficult. A clear goal is to achieve on-line feedback control of the process. This will require a much more detailed understanding of the process parameterization.

Diffusion Coating Methods

Diffusion coating is a surface modification process wherein the coating species is diffused into the substrate surface to form a protective layer. Diffusion coatings are the most used method for providing improved hot-corrosion and oxidation resistance for nickel-or cobalt-base superalloys. Diffusion coatings have been produced with aluminum, chromium, silicon, hafnium, zirconium, and yttrium alloys. The aluminum, chromium, aluminum-chromium duplex, aluminum-silicon duplex, and platinum-modified aluminide coatings are the most commercially significant. The surface of the superalloy is typically modified to a depth of 0.5 to 5.0 mils (thousandths of an inch) depending on the type of coating and the process parameters selected. These methods are sometimes classified as chemical vapor deposition methods. The committee considers three primary methods for producing diffusion coatings in this report: pack cementation, out-of-pack cementation, and chemical vapor deposition. Appendix E describes these processes.

COATING PROCESS CONTROL

Many of the recent advancements in coating technology have been the result of improved process control. For example, thermal spray technology has grown from an empirical art with highly variable results to a well-controlled process that gives consistently reliable results. Reliability has been achieved through improved control of carrier gas and power, control of feedstock material, and close monitoring of key process steps.

Process modeling, process monitoring, and real-time control must become more common if the manufacture of improved coatings is to be realized. Process control can only be achieved by understanding the empirical relationship between the coating process and the resulting coating. The choice of coating process should depend on a balanced approach among the technical attributes, the coatability issues, and the performance requirements demanded from the coated component. Developing the relationship of the process to product performance must be a priority, near-term endeavor for advanced coating systems. Such parametric data, guided by an understanding of coating behavior and failure modes, can be quickly and reliably acquired through rig testing, but the variability of the rig test must also be known. Rig-test modeling, parameter sensing, and feedback control will be an evolutionary and continuing near-term activity. The NRC (1989) study of *On-Line Control of Metal Processing* predicted this process evolution.

The current generation of metallic coatings, as well as the emerging TBC technologies, would benefit significantly from advances in process control. Both types of coatings are deposited by the same basic processes (i.e., plasma spray and physical vapor deposition), and process improvements will enhance the performance of both coating types. Improved on-line control must be developed to help ensure that the resultant behavior of the coated structure is highly reproducible and within the performance limits needed for the service requirements. For example, reducing the large variability in TBC performance and extending the service life are priorities. Many emerging nondestructive evaluation methods have yet to be effectively exploited (Murphy et al., 1993). A representative selection of methods are described in appendix C. The most promising nondestructive evaluation methods should be further developed and applied for manufacturing process control applications.

Accurate and comprehensive in situ measurements of coating processes are often difficult and sometimes impossible, leading to the need for process modeling. For example, in the typical direct-current plasma-spray system, powder particles are injected at approximately right angles into a high-velocity gas flame. The powder carrier gas increases turbulence within the flame and also disturbs its temperature distribution. Ideally, each particle becomes entrained in the flame, rapidly increases in temperature, melts (without excessive vaporization), and impacts the substrate where rapid solidification occurs. In practice, to formulate useful models, simplifying assumptions based on the underlying physical phenomena must be made and validated.

Within the manufacturing domain, the committee believes that important data do not currently exist in an appropriate form. The conceptual Taguchi Loss Function, outlined in figure 3-2, provides a starting point for developing a conceptual framework for identifying and classifying the data needed for a robust manufacturing process. The Taguchi Loss Function is based on the ideas of statistical process control and associates a deviation from the design-specified mean value of a parameter as a deviation from the minimum-cost performance product (Taguchi, 1993). A first step in applying this concept to coatings would be to chart the distribution of the parameters thought to be important to coating performance. Their frequency of deviations from the mean could then be determined. This would guide the search for those process parameters that must be controlled (i.e., those parameters that deviated most from the mean). Once a set of parameters needing monitoring and control are identified, process sensors (NRC, 1995b) and control systems can narrow the deviations from the mean. Appendix E contains more details of this approach.

SUMMARY

The majority of hot-section coatings are applied to protect superalloy components from degradation caused by the turbine engine environment. Since current hot-section structures are produced from nickel-and cobalt-base superalloys, substrate coatings in current engines have been optimized for superalloys. These coatings are primarily designed to protect the superalloys from oxidation and hot corrosion. Most recently, TBCs are being designed to reduce the effects of high temperatures and temperature gradients. Current and future generations of superalloys have extremely complex chemistries and microstructures, carefully tailored to meet the high demand of turbine

MATERIALS AND PROCESSES

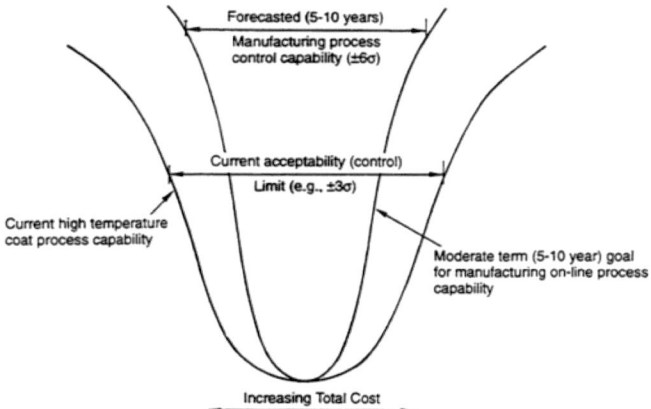

Figure 3-2 Schematic showing the benefit of development and deployment of manufacturing process control for high-temperature coatings.

engines. These complex substrate chemistries may be disturbed through interdiffusion with a coating, and their precisely crafted microstructures can be thrown into disarray as a result of coating process conditions. Therefore, coating/component combinations will have to be engineered together as a system to achieve optimum performance.

Thus the primary considerations in selecting a coating should be to protect the component and to ensure that the coating retains its protective qualities in the engine environment. Failure modes are discussed in chapter 4. Other considerations arise from various engineering factors, such as the requirement that the coating must also be compatible with the bulk material of the component under the demanding conditions in a turbine engine. These factors are further discussed in chapter 5.

Materials under development for use in future advanced components will also depend on coatings to protect against oxidation, hot corrosion, and embrittlement. Most of the new materials differ significantly from superalloys in physical, chemical, and mechanical properties. Also, many of these emerging materials, in their current form, are much less tolerant of flaws and failure in their coatings than the superalloys. Therefore, coating these materials presents significant challenges. Development of new coating materials and processes should be done concurrently with the development of the emerging materials.

4

Failure Modes

As discussed in chapters 2 and 3, most hot structures are fabricated from superalloys that have been tailored to meet the demands of turbine engine operation. Specific solutions to problems encountered under one set of operational conditions are often in conflict with those encountered under another set (Wood and Goldman, 1986). Thus, materials specialists have developed alloys that possess a balance of properties geared toward withstanding the most worrisome degradation mechanisms. As a result, alloying and processing of superalloys for high-temperature structural applications has historically focused on improving their creep resistance, with secondary goals of improving resistance to fatigue, oxidation, and hot corrosion (Sims et al., 1986).

This chapter reviews the failure modes experienced at high temperatures by current structural materials (i.e., superalloys) and their coatings. A case study of the use of failure modes to improve coating design is presented, using thermal barrier coatings (TBCs) as an example. The chapter concludes with a series of recommendations on future directions in research on environmental degradation modes.

DEGRADATION MECHANISMS OF STRUCTURAL MATERIALS

Degradation mechanisms for structural materials are a function of the engine operating conditions, engine mechanical design, and the component base materials. As a zero-order approximation, engine operating conditions determine operating temperature and structural loads, while engine design determines the amount of air made available to cool the hot structure. Table 4-1 summarizes the environmentally induced failure modes of interest (i.e., creep, fatigue, oxidation, and hot corrosion).

The first nickel-base alloys contained 20 percent chromium and successfully resisted oxidation within the operating temperatures then used because they formed a protective layer of chromia, Cr_2O_3. It was soon discovered that these alloys could be substantially strengthened with a coherent precipitate $Ni_3(Al, Ti)$, known as gamma-prime. However, the chromium content had to be reduced to increase the volume fraction of gamma-prime. While other elements could be added to make up for the loss in chromium, the alloys became more difficult to fabricate using existing methods (Stringer and Viswanathan, 1993).

As engine operating temperatures increased, chromia became less attractive as a protective scale because it oxidized further to the volatile compound of CrO_3. Alumina (Al_2O_3) remained attractive, however. Unlike chromia, alumina by itself does not protect against hot corrosion caused by molten alkali metal sulfates (Stringer and Viswanathan, 1993). Generally 4- to 5-weight-percent aluminum suffices for both gamma-prime and alumina formation.

Grain boundaries become a weak component of the microstructure at high temperatures. Creep deformation occurs in the grain boundaries, for example. To increase creep resistance, many of the critical hot-section airfoils are cast with the grains aligned parallel to the direction of the principal stress axis. These are called directionally solidified castings. Ideally, these components are selected as a single-crystal airfoil with the desired crystallographic orientation. Elements cannot segregate to the grain boundaries in single-crystal blades and cause brittleness. Thus, alloying flexibility exists to design a material with inherently high resistance to creep, oxidation, and hot corrosion. Single-crystal blades can be used uncoated in certain aeronautical turbine applications, but they do not yet possess sufficient inherent corrosion resistance for industrial turbine use, which is much more demanding on service life (Stringer and Viswanathan, 1993).

The degradation modes common to cooled hot-section superalloy components include low-cycle thermal fatigue, oxidation, and creep, with creep being the least important factor because it is designed out. Combustor and turbine blade components also degrade during high-cycle fatigue processes. On the other hand, noncoated components, such as low-pressure turbine blades, are often creep limited.

DEGRADATION MECHANISMS OF COATINGS

Once it became apparent that it would be extremely difficult to develop structural materials that possessed all the desired high-temperature mechanical properties as well as environmental resistance, the functions were uncoupled. Structural alloys were developed to optimize mechanical properties, and coatings were developed to serve as physical

TABLE 4-1 Environmentally Induced High-Temperature Structural Material Failure Modes

Failure Mode	Definition
Creep	Time-dependent, thermally activated inelastic deformation of a material. The rate of creep increases as the temperature increases for constant stress.
High-Cycle Fatigue	Microstructural damage mechanism that results from small stress amplitude cyclic loading, such as vibrations; failure will occur after a relatively large number of cycles.
Low-Cycle Fatigue	Microstructural damage mechanism that results from large stress amplitude cyclic loading; failure will occur after a relatively small number of cycles. Very abrupt thermal changes, such as from engine start and stop cycles, are the driving force for this failure mode.
High-Temperature Oxidation	Solid-gas chemical reaction that produces the oxide(s) of constituents within the solid. The rate of oxidation increases exponentially with temperature, certain oxides (notably those of aluminum and chromium) are slow growing and protective of the underlying substrate.
Hot Corrosion	Electrochemical reaction between substrate and molten salts, typically sodium and potassium sulfates. Two forms of hot corrosion are generally recognized: Type I (high temperature), which typically occurs between the temperatures of 820°C and 920°C (1500°F and 1700°F), with a maximum at about 870°C (1600°F), characterized by the buildup of a nonprotective oxide layer as oxidation and sulfidation destroy the metal substrate; and Type II (low-temperature), which typically occurs between 590°C and 820°C (1100°F and 1500°F), with a maximum at about 700°C (1300°F), often exhibiting pitting.

barriers between aggressive environments and the substrate (Stringer and Viswanathan, 1993).

The structural materials degradation modes that are moderated by the traditional metallic protective coatings include oxidation and hot corrosion. In addition, TBCs insulate the underlying structure against the full effect of gas-path heat. They thus retard creep degradation and reduce the severity of the thermal gradients and transients in the structure that drive low-cycle fatigue processes. By dampening the amplitude of vibrations, some coatings can reduce high-cycle fatigue. However, no such coating to date has been found that can survive the aggressive turbine environment.

Table 4-2 summarizes the various failure modes that pertain to the coatings themselves. Historically, metal coatings (i.e., alumide, chromide, and MCrAlY) have been designed to withstand three types of environmental attack: high-temperature oxidation, high-temperature (Type I) hot corrosion, and low-temperature (Type II) hot corrosion. Figure 4-1 shows the range of temperatures over which these attacks occur. However, it should be kept in mind that the oxidation of the aluminide and chromide coatings at the coating/gas-path interface results in the formation of a protective oxide scale. Therefore, within this context, high-temperature oxidation is not purely a degradation mechanism. In addition to these failure modes, thermomechanical fatigue of coatings (and substrates) can occur as a result of cyclic and thermal loading of the component. Thermomechanical fatigue cracking in coatings, particularly around film-cooling holes, has often been observed in advanced engines. A TBC can reduce the magnitude of the thermomechanical fatigue strain range and can minimize or eliminate thermomechanical fatigue cracking.

Several other damage modes cause coating loss and accelerate the overall failure mechanism:[1]

- mechanical distress to the coating, such as nicks and gouges, which is caused by objects ingested into the engine airstream
- solid-state diffusion of elements between coating and substrate, which can lead to the loss of critical elements from the coating and formation of undesirable phases in the substrate
- spallation caused by differential thermal expansion between the coating and the substrate, which can lead to mechanical failure of the coating
- rumpling of metallic coatings as a result of creep

The coating degrades at two fronts during service: the coating/gas-path interface and the coating/substrate interface. At temperatures well below the incipient melting point of conventional superalloys, deterioration of the coating surface at the coating/gas-path interface tends to be a consequence of oxidation or hot corrosion. As the temperature rises, diffusion across the coating/substrate interface plays a greater role in

---

[1] These damage modes can be addressed by proper design of the coating system (as discussed in chapter 5) and operational procedures.

TABLE 4-2 Environmentally Induced High-Temperature Coating Failure Modes

Failure Mode	Definition
High-Temperature Oxidation	Solid-gas chemical reaction that produces the oxide(s) of constituents within the solid. The rate of oxidation increases exponentially with temperature; certain oxides (notably those of aluminum and chromium) are slow growing and protective of the underlying substrate.
Hot Corrosion	Electrochemical reaction between metal and molten salts, typically sodium and potassium sulfates. Two forms of hot corrosion are generally recognized: Type I (high-temperature), which typically occurs between the temperatures of 820 and 920°C (1500°F and 1700°F), with a maximum at about 870°C (1600°F); and Type II (low-temperature), which typically occurs between 590 and 820°C (1100 and 1500°F), with a maximum at about 700°C (1300°F).
Mechanical Distress	Erosion and impact damage caused by the ingestion of particles in the air stream.
Solid-State Diffusion	Reduction of the aluminum content of the coating because of interdiffusion with the base metal.
Spallation	Loss of protective oxide at the coating/oxide interface.
Thermomechanical Fatigue Cracking	Long-term (i.e., over many cycles) formation and propagation of cracks because of external mechanical stresses and to residual stresses from lack of thermal expansion compatibility between the substrate and the coating

degradation; compositional changes at this interface can also compromise the structural properties of the substrate.

The remainder of this section reviews the key characteristics of coating degradation and suggests areas in which additional knowledge is needed. The discussion is limited to commercial coatings for which service-derived experience is available.

High-Temperature Oxidation

At high temperatures, coatings that protect against oxidation form a compact, adherent oxide scale (usually Al_2O_3) that provides a barrier between the high-temperature gases and the underlying metal. Chromia scales (Cr_2O_3) have been used but offer less protection than alumina above 840-870°C (15501600°F) because chromia scale tends to sublimate to CrO_3 above these temperatures. Without the protective scale, the coating, and ultimately the substrate, come under rapid attack.

The general case is that the oxide spalls repeatedly until it can no longer form. Then internal oxidation, caused by diffusion of oxygen into the coating, degrades the protective oxide scale. Figure 4-2 shows the result of high-temperature oxidation of a coating and its penetration to the base metal. In this example, an adherent, relatively uniform external oxide scale remains, but internal oxidation of the coating and base metal has occurred. Recent work (Smialek, 1991; Smialek and Tubbs, 1995) has demonstrated the enhancements to oxide adherence that can be achieved as a result of very tight control of contaminant levels (e.g., maintaining sulfur levels in the < 1 ppm range). Segregation at the metal/scale interface causes disruption of the bond and spallation of the oxide. In addition, oxygen-active elements such as hafnium and yttria are often added to coatings to improve the

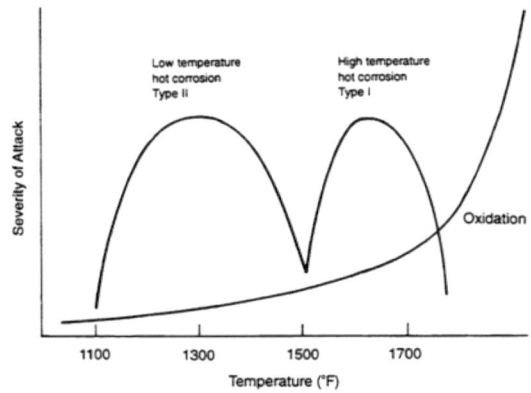

Figure 4-1 Types of high-temperature attack for metallic coatings (aluminide, chromide, MCrAlY, etc.) on nickel-base superalloys with approximate temperature regimes and severity of attack. Source: EPRI (1991). Copyright 1991. Electric Power Research Institute. EPRI GS-7334-L. *Guidebook and Software for Specifying High-Temperature Coatings for Combustion Turbines.* Reprinted with permission.

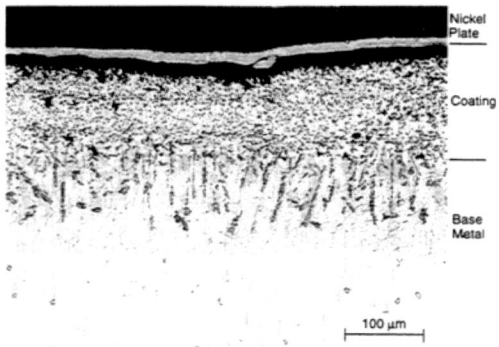

Figure 4-2 Micrograph of service-exposed CoCrAlY overlay coatings showing internal oxidation of coating and base metal. Source EPRI (1991). Copyright 1991. Electric Power Research Institute. EPRI GS-7334-L. *Guidebook and Software for Specifying High-Temperature Coatings for Combustion Turbines.* Reprinted with permission.

adhesion of the oxide. The mechanisms proposed for the "reactive element" effect on scale adhesion include (1) reducing the activity and mobility of sulfur, (2) mechanical pegging, (3) reducing void formation, (4) increasing scale plasticity, and (5) oxide doping.

The fundamental oxide-scale formation, growth, and breakdown have been studied for several years. Although some unresolved issues linger, high-temperature oxidation is relatively well understood. Less clear are the effects of oxidation from secondary species in the fuel and combustion air. The role of some of these contaminants have been studied; for instance, it is known that chlorine accelerates the rate of oxidation. However, a quantitative understanding of these effects is not well developed.

Hot Corrosion

Hot corrosion involves attack by molten salts, typically sodium and potassium sulfates, that enter the turbine hot section as contaminants from the air and fuel and can result in rapid loss of material (Stringer and Viswanathan, 1990). As indicated in table 4-2, there are two forms of hot corrosion: high-temperature (Type I) and low-temperature (Type II). The basic mechanisms of hot corrosion are relatively well understood, although there still remain areas of disagreement (Rapp and Zhang, 1994). The role of secondary elements (e.g., impurities such as chlorine, calcium, and iron from the local environment) is not as clear.

The key to continued protection against hot corrosion is maintaining the protective scale and keeping the hot-corrosion process in its initial, or incubation, period. Rapid loss of the coating and the underlying metal substrate will occur once the scale is lost or penetrated. The ideal protective scale would minimize solubility of the fused salt, a property determined by the combustion chemistry and operational environment. This criterion for petroleum-base fuels can generally be satisfied by either chromia or alumina, although alumina protects at higher operating temperatures than chromia.

A TBC degradation mechanism has been observed in service in which molten surface deposits can accelerate spalling of electron-beam physical vapor deposition (EB-PVD) and plasma-sprayed TBCs. Salt species in vapors from such sources as dirty fuel or marine environments can condense onto the coating surface. If the coating temperature is above the melting point of the salt, it can wick through the interconnected cracks and porosity of the coating (Miller, 1986). The wicked-in salt freezes on cooling, eliminating the strain tolerance mechanism of the coating and accelerating spallation (Palko et al., 1978; Miller, 1986; Strangman, 1990).

Mechanical Distress

Two related failure modes-mechanically induced erosion and impact damage-merit special consideration. Erosion results from particles in the air stream that abrade the coating and accelerate its loss. Impact damage occurs when a solid object in the air stream strikes the coating. These failure modes are sometimes overlooked because they can often be avoided by proper operating practice. However, these modes will become increasingly important as substrate materials are operated under conditions in which they offer little inherent protection against oxidation or other major failure modes. These failure modes are also a key concern in the retention of TBCs. These ceramic coatings are inherently more vulnerable to impact damage and erosion than metallic coatings.

Creep of stand-alone metallic coatings at higher temperatures can cause rumpling of the surface of coated components resulting in degradation of performance and durability. Application of a TBC may alleviate the problem by lowering the temperature of the metallic coating (Bose and DeMasi-Marcin, 1995). However, plastic instability of bondcoats for TBCs is still a potential problem at higher temperatures, and work on creep-resistant bondcoats is continuing (Brindley, 1995).

Solid-State Diffusion

The loss of aluminum in the coating, caused by diffusion of the coating with the base metal, ranks as another important degradation mode. Diffusion of aluminum into the base metal and base-metal elements into the coating reduce the concentration of aluminum in the coating that is available for forming alumina. A protective oxide can no longer re-form after spallation once the aluminum concentrations fall below a certain level. The basic concepts of diffusion are well understood for simple systems, and the diffusion of complex, multielement systems containing multiple phases can be formally described (Nesbitt and Heckel, 1984). Obtaining actual interdiffusion coefficients and predicting the interdiffusion in these systems are formidable tasks, however. Consequently, reliance on empirical measurements of interdiffusion is necessary and usually sufficient for engineering purposes.

Spallation

Loss of oxide scale by spallation is the major concern for all coatings and poses a particular problem for coatings based on aluminum or chromium. Once again, a protective scale can no longer be maintained when the aluminum (or chromium) concentration falls below a critical level, usually cited as approximately 4- to 5-weight-percent aluminum.

Loss of the protective oxide at the coating/oxide interface is most damaging since an entirely new protective scale must be formed. Recent research indicates that spallation at this interface can be exacerbated by the presence of sulfur in the material. With the reduction of sulfur impurities, stepwise improvements have been observed in the adherence of protective oxides to superalloys (Smialek, 1991; Smialek and Tubbs, 1995), resulting in vastly improved oxidation resistance. Although studies have not completely elucidated the role of sulfur, lower levels of sulfur in both the superalloy base and its coating increase the coating's resistance to cyclic spallation.

TBC thickness is yet another concern. While it is obvious that a finite amount of TBC must be deposited in order to reap the benefit of a thermal barrier, increasing TBC thickness adds to component weight and accelerates spallation (Bose and DeMasi-Marcin, 1995).

Spallation of TBCs is an important failure mode. These ceramic coatings are not self-replenishing. Hence, spallation rapidly degrades the insulating properties of the coating and can accelerate attack of the underlying metallic bondcoat or substrate. This weakness of TBCs is discussed in greater detail in the next section.

A CASE STUDY: DEGRADATION OF THERMAL BARRIER COATINGS

Experimental and theoretical thermal characterization of TBCs has shown that they can provide significant thermal benefits to components in operation, but only if coating integrity is maintained. TBCs are applied on top of a MCrAlY coating; the TBC provides thermal insulation, and the MCrAlY protects the substrate against oxidation and hot corrosion. Zirconia is extensively used as a TBC since it has low thermal conductivity and a relatively high coefficient of thermal expansion. The zirconia used is partially stabilized in the metastable tetragonal phase with 6- to 8-weight-percent yttria. TBC coatings are applied using one of two processes: plasma spray and EB-PVD, as discussed in chapter 3.

Maintaining coating integrity for a TBC is more difficult than for metallic coatings because of the relatively large change in properties from the metallic substrate to the ceramic coating. The problem is exacerbated by the fact that TBCs provide the greatest efficiency benefit at the highest operating temperatures, where TBC failure is most likely to cause unacceptable component damage. Thus, while coatings with even lower conductivity may be possible (see chapter 8), the high variability of TBC thermal cycle life suggests that the use of these coatings to their full potential will depend on their susceptibility to failure.

The microstructure of the coating resulting from plasma spray differs significantly from that created by the EB-PVD processes (see figure 4-3). Hence the coating failure modes are not the same. Unlike the other coating types that may become ineffective through chemical degradation while the coating is still intact, the loss of the effective life of a TBC is

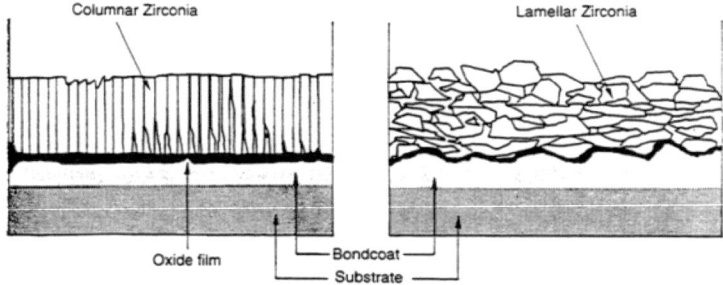

Figure 4-3 Comparison of the microstructure of EB-PVD and plasma-sprayed TBCs.

usually marked by the physical loss of the ceramic insulating layer. This loss occurs by one of two means: ceramic layer spallation or erosion of the ceramic layer.

Spallation

TBC spallation is a mechanical event that results in the removal of the ceramic layer by delamination from the bondcoat. Numerous processes may trigger this mechanical event, but the two main culprits are bondcoat oxidation and the strain generated by a thermal expansion mismatch between the ceramic top coat and the metallic components of the system (Miller and Lowell, 1982; Strangman, 1985; Hillery et al., 1988; Demasi-Marcin et al., 1989).

Data indicate that oxidation of the bondcoat and thermal expansion mismatch strains between the ceramic top coat and the metallic bondcoat have a synergistic effect on TBC failure. Under realistic conditions, the action of either alone cannot cause failure of plasma-sprayed (Miller and Lowell, 1982) or EB-PVD TBCs (Manning-Meier et al., 1991). This generality may not hold for conditions of extreme thermal transients, however, such as for rocket engine applications (Brindley and Nesbitt, 1988). Since oxidation and thermal-cycle strains must both be present to cause TBC failure, reducing either could dramatically improve TBC life. However, the details of degradation differ for EB-PVD and plasma-sprayed TBCs because of their substantially different ceramic layer and interface structures.

Plasma-Sprayed TBCs

Figure 4-4 shows the microstructure of a typical plasma-sprayed TBC. Failure of a plasma-sprayed TBC generally consists of cracking and spallation. Delamination cracking generally results from high in-plane compressive stresses that cause significant out-of-plane tensile stresses and produce delamination cracks parallel to the interface and near the peaks of the rough bondcoat. Finite-element analysis of a realistic rough interface of a plasma-sprayed TBC has shown that the thermal-cycle stresses in the ceramic layer at the bondcoat peaks are roughly two orders of magnitude higher than expected at a smooth interface (Ferguson et al., 1994). Thus the rough interface of a plasma-sprayed coating not only increases initial adherence with the ceramic layer through mechanical interlocking but also drives its delamination and reduction in life. Studies in which surface roughness was varied using an analytical (Evans et al., 1983) or numerical model (Ferguson et al., 1994) have identified the radius of curvature of the bondcoat peaks as the single most important feature in generating stress. Spacing and amplitude may also contribute.

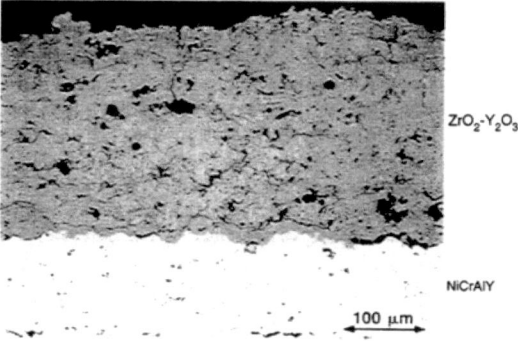

Figure 4-4 Photomicrograph of a plasma-sprayed TBC.

Out-of-plane cracks are also present in plasma-sprayed TBCs, but these result from processing and in-plane tensions applied to the ceramic layer during heating of the component. The out-of-plane cracks are not generally believed to be a problem for coating durability and could even enhance the strain tolerance of the coating. The out-of-plane cracks could play a role in spallation if they were to link major horizontal delamination cracks, and, in fact, out-of-plane cracks have been shown to accelerate spallation in thicker coatings (Bose and DeMasi-Marcin, 1995). These cracks could also reduce coating life by adding to the in-plane compressive stresses of the ceramic layer after cooling from high temperature. This could occur if the cracks were to open during heating but not close on cooling because of crack-face interference or cracking debris lodged in the crack.

Delamination in a plasma-sprayed TBC does not occur with a single cracking event but by progressive crack growth (demise et al., 1989; Cruse et al., 1992). Since cracks, flat pores, and abundant potential crack growth paths along splat boundaries are present in the as-processed structure, delamination cracking requires only crack growth and linkup through thermal fatigue. No crack initiation step is required. Thus, long service lives are possible if the strain imposed on the ceramic layer during thermal cycling is low enough. This requires a reduction in the strain imposed on the ceramic layer by thermal expansion mismatch, oxidation, or both.

EB-PVD TBCs

EB-PVD coatings tend to fail at the alumina-scale/bondcoat interface. Finite-element analysis modeling of the thermal cycle behavior of PVD coatings indicates that out-of-plane tensile stresses are well below the initially high adhesion strength of the alumina and even below the adhesion strength of coatings that have been thermally cycled (Manning-Meier et al., 1991). These low stresses primarily result

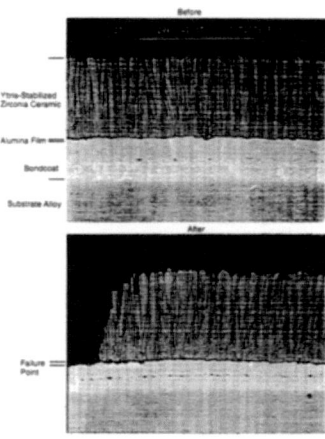

Figure 4-5 Photomicrographs of EB-PVD TBCs before and after failure.

from the smoothness of the interface between the bondcoat and the alumina layer. Figure 4-5 shows an EB-PVD coating before and after failure.

Although the calculated out-of-plane tensile stresses are lower than the adhesion strength of the coating, Manning-Meier et al. (1991) proposed that alumina delamination proceeds by progressive interfacial cracking from in-plane strains. Since the failure occurs at the alumina/bondcoat interface, progressive reduction of the adhesion of the alumina layer may also be involved in spallation of the TBC.

Oxidation is a larger fraction of the failure driver for PVD coatings than for plasma-spray coatings. The difference apparently arises because PVD coatings are less susceptible to thermal cycling damage (Manning-Meier et al., 1991). Continued growth of the oxide scale is believed to increase the compressive stresses applied to the ceramic layer during thermal cycling (Rigney et al., 1995). Several other mechanisms may increase the mechanical strain on the ceramic layer: sintering (Eaton and Novak, 1987; Manning-Meier et al., 1991), formation of high-volume or low-strength oxides at the bondcoat/ceramic interface (Lee and Sisson, 1994), and creep of the bondcoat (Hillery et al., 1988; Brindley and Whittenberger, 1993).

Erosion

TBCs are more susceptible to erosion than fully dense ceramic coatings because the strain-tolerant structure of TBCs includes crack like features. Furthermore, the relatively fine microstructure of the TBC coating means that the distance from one crack like feature to another is small and the crack length required to remove small portions of the coating is also quite small. Thus, the erosion rate, even compared with other porous ceramics, is expected to be high.

The degree of erosion depends on the density of particulates in the working gases and the angle of impingement. Particulate loading is expected to be higher for engines operating in dusty environments (e.g., aircraft engines ingesting debris during takeoff and landing and land-based or aircraft turbines in dusty sites or geographic regions such as the Middle East). Erosion can also stem from the particulates generated inside the engine (e.g., coking or salt shedding from parts that have deposited condensates). In all cases, erosion is more significant on the leading and trailing edges of rotating components than on static components or elsewhere on turbine blades.

Erosion removes the ceramic layer of the TBC and thereby removes the insulation it provides. Current practice is to design the leading edges of turbine blades as if no TBC were to be applied. Thus the heat load on the component will not exceed the design temperature for the leading edge if the TBC were eroded. As operating temperatures increase and the drive for higher efficiency continues, a solution to the leading-edge erosion problem will become increasingly important. Fabricating thicker TBCs at high-erosion areas is not an attractive solution because of the added weight of the coating and the increased likelihood of cooling hole blockage. A more satisfactory solution would be the development of an erosion-resistant coating or surface layer for current TBC materials.

RESEARCH OPPORTUNITIES

Engineers have a good understanding of how high-temperature oxides form. The role of minor elements (e.g., sulfur) in high-temperature oxidation and composition needs additional study, as do spallation of the oxide during both steady-state operation and thermal cycling.

The mechanisms of both low-and high-temperature hot corrosion are also relatively well understood. The role of secondary elements in hot corrosion needs additional study, as does the effect of mechanical integrity of the oxide scale during hot corrosion. In order to develop realistic fuel specifications, the resistance of coatings to hot corrosion in biomass and gasification environments (i.e., low sulfur but high alkali) needs clarification.

Research is required to improve and apply models that can help optimize coating development and to provide more accurate estimates of the remaining coating life for in-service engines. More specifically, advances are required in (1) the understanding of the failure mechanisms and factors involved as a function of use conditions; and (2) the development of qualitative and quantitative life-prediction models based on all active failure mechanisms.

5

Engineering Considerations

Choosing a high-temperature coating for a turbine engine is currently a sequential design process that is dictated by the operating conditions of the engine. The most suitable combination of substrate cooling schemes and alloys are selected for the hot-section components. Coating systems are then specified if they are needed to extend the performance of the components to higher temperature regimes, to improve the durability of components, or to allow the use of fuels with a broad range of thermal efficiencies and contaminant levels.

In addition to evaluating the coating material on its own merits, the designer must take into account several other engineering variables to select the best high-temperature coating system. For example, coatings are typically designed to wear out (i.e., lose their dimensional tolerance) without causing premature component replacement or unpredictable or catastrophic structural failure. The hope is, of course, that a coating will improve all aspects of component performance, from environmental resistance to fatigue resistance. This is not usually the case, however. Components generally perform better with the coating but not as well as they could if the environmental degradation driver, and thus the coating, could be eliminated. An inappropriate coating can also actually reduce performance. Thus the advantages of applying a coating must always be weighed against the disadvantages (Wood and Goldman, 1986; Dyson, 1989).

This chapter discusses the key engineering factors that must be evaluated in specifying a coating system and, where appropriate, emphasizes the shortcomings of the traditional, sequential design process described above.

COMPATIBILITY OF COATINGS WITH STRUCTURAL MATERIALS

The coating system must be compatible with the component's base material in order to protect against environmental attack and to prevent degradation caused by the coating itself. Three different types of compatibility are generally considered important: chemical (metallurgical), processing, and mechanical.

Chemical (metallurgical) Compatibility

Many coatings require some interdiffusion with the substrate to ensure adhesion (e.g., diffused aluminide and MCrAlY-type overlay coatings), but these coatings must be relatively stable to avoid excessive coating/component interdiffusion and chemical reaction. Simple interdiffusion can erode critical elements (e.g., the aluminum used to form the alumina scale; Nesbitt and Heckel, 1984) and degrade the coating's protective properties. Interdiffusion with the structural alloy can create deleterious reaction zones near the interface. Brittle intermetallic compounds and other phases often precipitate in these zones, leading to cracking near the interface and spallation of the coating. A brittle reaction zone may also reduce component ductility and fatigue resistance. In such an event, the coating may *decrease* component life. Other detrimental effects include the formation of low melting-point (eutectic) phases and low-strength regions in the substrate material because of unwanted changes in the local chemistry of the alloy.

Walston et al. (1993) cited a case in which a coating decreased the performance of a new superalloy. The research goal was to develop a nickel-base alloy with higher creep strength. The work indeed demonstrated improved creep properties for an uncoated specimen, but the application required that the component be coated to protect against oxidation. When an aluminide coating was applied, this alloy had much lower strength and was more vulnerable to stress rupture. This poor performance was attributed to the formation of brittle phases near the interface due to a coating/substrate reaction. The beneficial properties of the new alloy were restored after a slight modification to the chemistry of the base material eliminated the formation of the brittle phase (Walston et al., 1993). This case illustrates the sensitivity of carefully optimized superalloy compositions to slight changes in chemistry. It is also one of the few reported cases in which a substrate composition was changed to be more compatible with a coating.

Process Compatibility

The coating material may be completely compatible with the component but the coating process may be incompatible. This can occur, for example, when the coating process requires high temperatures or a special precoating surface treatment that changes the properties of the structural material. Incompatibility also arises if the process cannot apply a coating to the required geometric form, reach all areas of the component (such as internal passages), provide the necessary surface finish, or apply different thicknesses of coatings on a part, as is sometimes required.

For example, chemical vapor and physical vapor deposition processes require high temperatures to grow the correct coating microstructure and ensure that the coating adheres to the substrate (Lammermann and Kienal, 1991). These temperatures may impair the structural material or cause unacceptable dimensional changes in the component, such as warping. Applying heat or mechanical straightening after coating may restore desired properties and dimensions, but restoration may not be possible in some cases.

Some coatings require a particular surface treatment before they can be applied. Unfortunately, some pretreatments may significantly damage the substrate. For example, thermal spray coating usually requires pretreatment with grit blasting to roughen the surface and improve adherence. Grit blasting of a single-crystal vane or turbine blade should generally be avoided, however, since it induces residual strains that can serve as nucleation sites for recrystallization during subsequent operation of the turbine.

Mechanical Compatibility

To maintain the protective features of the coating, the mechanical properties of the coating must be designed to accommodate or match those of the substrate. Factors affecting mechanical compatibility include the coefficient of thermal expansion (CTE), coating ductility, parasitic weight, cohesion, adhesion, and surface roughness. To ensure mechanical compatibility, the coating must have some strain tolerance to accommodate any such mismatches.

Maintaining the protective features of oxidation and hot-corrosion-resistant coatings generally depends on maintaining adherence and a microstructure free of through-thickness cracks or other defects. For TBCs that are precracked to help accommodate strain, continued protection generally requires that the coatings remain adherent and insulated over bondcoats that are typically oxidation-resistant coatings.

Coating materials and processing should maximize the longevity of the component (i.e., coatings are selected with properties that generate the lowest stresses in the coating/substrate system). One consideration is the coating/component CTE match. Matching the CTE of the coating and the substrate is an extremely difficult process. If the coating and substrate CTEs match relatively closely, the stresses generated during processing and thermal cycling should be small. Under these conditions, even a coating with low ductility will not fail catastrophically (i.e., crack), allowing it to remain adherent and offer protection. However, if there is a large CTE mismatch, even a ductile coating may crack or spall due to the large stresses and strains incurred during thermal cycling that result from thermal expansion mismatch. Another consideration is environmentally enhanced thermomechanical fatigue, which is a special case of low-cycle fatigue that is not adequately predicted with isothermal low-cycle fatigue data and models. Life-prediction methods are needed that help designers avoid thermomechanical fatigue cracking of coated superalloy components.

All coating applications require some degree of strain tolerance. Because most coatings have less ductility than their substrates, coated components have lower ductility. As a consequence, the strain required to initiate a crack in the structure may prove less than that in an uncoated specimen. For example, a crack that originates in a coating (i.e. caused by thermomechanical cycling, etc.) can trigger cracking in the base material, with a subsequent loss in ductility and fatigue life of the component (Smialek et al., 1990).

Coatings that protect against oxidation or hot corrosion must accommodate strain and have enough ductility to keep from cracking under operating conditions. For TBCs, the low effective modulus imparted by discontinuities in the coating structure usually provides the strain tolerance. This tolerance allows the coating to deform locally without generating the high stresses that would cause coating spallation.

There is often a subtle tradeoff between improved environmental performance and component life. While coatings can improve environmental resistance, their extra mass increases stress on a rotating part. This in turn decreases the creep life and adds to the stress imposed by thermal gradients. Furthermore, the extra mass on blades also adds to the load on the turbine disk as well as on the turbine shaft and its bearings. Therefore, the entire rotating system must be designed to handle the additional weight of the coatings. If not, engine life will be severely compromised.

Component degradation aside, coatings directly affect component function. For example, a coating changes part dimensions, which can be very significant for thin sections. A coating can thicken the trailing edges of turbine blades and vanes, causing turbulence and a loss in aerodynamic efficiency. Even components designed to allow for the added thickness of the coating suffer mechanical degradation because of the higher trailing-edge thicknesses. Coating the leading and trailing edges, which are important to aerodynamic efficiency, prove especially challenging because of the combination of mechanical and thermal strains coupled with the erosion that may occur at these highly dynamic locations.

Coatings can also change the surface finish of a component, affecting both aerodynamic drag and heat transfer rates. For example, a recent experimental and theoretical analysis of the Space Shuttle's main engine turbopump turbine section compared performance for two cases: turbine blades with a rough 400-micro-inch finish (R_a) and blades with a smooth 30-micro-inch finish (Ra; Boynton et al., 1992). The analysis indicated that the efficiency of the turbine increased by an absolute value of 2.5 percent (not 2.5 percent of the baseline) when the coating was smoothed. Thus, while the change in efficiency varies with engine operating conditions, there was a significant advantage to a smoother surface finish.[1]

While smooth surfaces improve aerodynamic performance, controversy exists regarding the type of surface finish that should be required for engine components, particularly airfoils. These components accumulate deposits and erode, which may make costly surface finishing operations meaningless early in their service life. For some applications, original surface finish is important, but researchers disagree on how to specify the finish. At present, surface finish is usually specified as the arithmetic average roughness, a measure termed Ra. The value of Ra can be identical for surfaces with very different surface profiles and potentially different aerodynamic responses. Thus, this measure alone cannot determine the aerodynamic quality of the surface. A thorough experimental and theoretical treatment of roughness effects would benefit the coatings community.

Coating surface roughness may also affect the heat transfer rate. In particular, surface roughness generally increases the heat transfer rate under laminar flow conditions. However, the effect of increasing roughness under turbulent conditions is less clear. Since both component and coating degradation are generally sensitive to temperature, adding heat to a coating can degrade the component. The few experimental studies conducted under these conditions have yielded conflicting data (Boyle and Civinskas, 1991). Roughness effects on heat transfer require continued investigation.

COMPONENT COATABILITY

Coatability is the ability of a coating process to deposit a coating on the required surface. Expressed in these terms, coatability is a function of geometry and size rather than base material. A primary issue is whether the process can coat all of the surfaces required. As discussed in chapter 3, many coating processes cannot coat surfaces that are not in the line-of-sight (e.g., the internal passages of airfoils and sharp changes in surface geometry). If a component is large enough, physical manipulation may expose hidden surfaces to the coating source.

Uniform coating thickness is another aspect of component coatability. Most coating processes tend to produce variations in thickness at edges, inside corners, and on irregular contours. This variation may impair the coating or reduce component life. One obvious solution is to exclude processes that foster nonuniformity. Alternatively, it may sometimes be possible to refinish a coating (i.e., remove regions that are too thick) to achieve a uniform thickness.

Components containing cooling holes affect the selection of the coating process. Cooling holes are often present prior to coating. During application, the coating may partially block the holes, a phenomenon known as coat-down (i.e., coating is deposited on the inside surfaces of the hole). Since the holes can be small (e.g., 0.5 mm in diameter), coat-down can block a significant fraction of the hole, diminishing cooling to the outside surfaces. The preferred means for maintaining adequate hole size is to use a coating method that limits the amount of coat-down. Other methods, such as laser drilling of holes after coating, can also be used but are more costly.

The size of the component is another important consideration. Many current processes use enclosed tanks or reactors (e.g., electron-beam physical vapor deposition, low-pressure plasma spray, metalorganic chemical vapor deposition, and pack-chemical vapor deposition processing). Large parts, such as turbine blades or combustors for large-frame power-generation engines, may not fit into some of these processing chambers. It may also be difficult for the processes that were initially designed for small parts to achieve the same coating uniformity and coating structure for large parts. Such difficulties may arise in maintaining uniformly high substrate temperatures, which are required for some processes. Large parts may also be difficult to manipulate in such a way as to allow all of the surfaces to be coated.

OTHER ENGINEERING CONSIDERATIONS

Coating Databases

Traditionally, high-temperature coating data are obtained only after the mechanical properties of the uncoated substrate have been well characterized. These data are generally specific to the application required and are often proprietary. Moreover, the general utility of the data is limited because high-temperature coatings are applied by a variety of processes with a variety of controls.

---

[1] Since some coating processes, such as thermal spraying, lead to high surface roughness, surface-finishing methods have been developed for these processes. There is also a tendency to use coating processes that inherently provide smooth surfaces, such as electron-beam physical vapor deposition and chemical vapor deposition.

Properly structured databases could serve as effective mechanisms for the collection, organization, and dissemination of data and experience for new coating technologies (NRC, 1995c). For example, the design criteria for TBCs can become more sophisticated if the existing data gathered from previous laboratory experience is made available in an organized fashion to design teams. Laboratory standards are also needed to identify and characterize the properties of materials (and coatings) over very long (50,000+ hours) service life requirements. Standards for coatings are discussed in appendix A.

Management of Air, Fuel, and Water

Contaminants found in the fluids required to operate a gas turbine (air, fuel, or water or steam[2]) can combine in the hot section to produce corrosion, erosion, and deposition under certain temperature and pressure conditions. For land-based turbines, airborne contamination arises from particles and gases from the ambient air, from evaporative cooling of the incoming air (for power augmentation), and from on-line engine cleaning. Regardless of the source and means by which contaminants enter the gas path, they accelerate degradation of high-temperature components beyond that dictated by the temperature at the component surface.

Gas turbines burn a variety of fuels, the power of which vary in form and content. The early literature was mostly concerned with petroleum-base fuels, but the development of other fuels (e.g., coal and coal-derived fuels) has been the subject of programs sponsored by the U.S. Department of Energy (DOE, 1992a,b; NRC, 1995a).

The petroleum-base liquid fuel used in aircraft engines is strictly controlled so that indigent fuel contaminants do not damage the engine. Enabling gas turbines to operate on lower quality liquid fuels (i.e., residual fuels) without accelerating deterioration from contaminants remains a challenge to the development of hot-section coatings. In contrast, most power generating units run on natural gas and are therefore expected to suffer less degradation from fuel combustion products.

It is standard practice for turbine design engineers to specify maximum concentrations of harmful contaminants. Addressing the combined concentration of contaminants in air, fuel, and water has proven effective in managing corrosion and deposition for land-based turbines. This methodology involves calculating concentrations in the turbine section relative to allowable limits. The amount that the contaminants will need to be reduced can then be calculated.

Figure 5-1 depicts the tradeoff between the initial investment in fluid cleanup and the cost of power for industrial uses. Capital investment is directly correlated to the quality of the gas-turbine environment (represented in this case by the concentration of alkali metals in the combustor effluent). This investment is offset by the cost of maintenance, including turbine airfoil replacement or repair as well as the longer mean time between overhauls (Hsu, 1987, 1988).

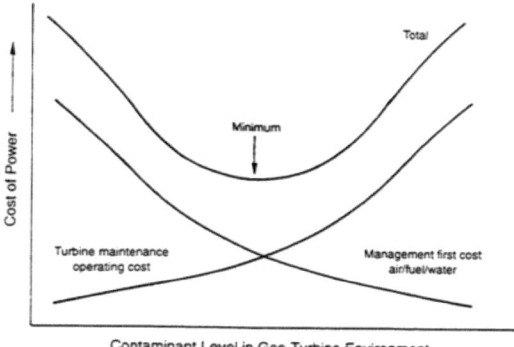

Figure 5-1 Tradeoffs between first cost and operating cost.

Influence of Coatings on Gas-Turbine Emissions

Gas turbines can produce harmful emissions during the combustion process. These emissions include oxides of nitrogen (NO and NO_2, commonly called NO_X), carbon monoxide (CO), unburnt hydrocarbons, sulfur oxides (mainly SO_2 and SO_3), and particulate matter. The character of these emissions depends on a number of factors: the temperatures in the combustion zone; the mixture of air and fuel in the combustor; the residence time, particularly in the hotter parts of the flame; and the nature of the fuel. In general, a gas turbine operates with much more air than is required for the stoichiometric combustion of the fuel. This air is introduced in different parts of the combustor, however. The actual air-to-fuel ratio may be close to stoichiometric at the burner itself; secondary air is then mixed in later, and the combustor is designed to ensure good mixing between this and the primary combustion products. Further air is added before the combustion gases enter the first row of stationary vanes.

Poor mixing is evidenced by the passage of unburnt hydrocarbons through the turbine.[3] This often happens during startup, when the fuel-to-air ratio is high; the smell of unburnt hydrocarbons is usually detectable at airports. Incomplete combustion of the carbon can also result in the emission of CO from the combustor, which may not oxidize to CO_2

---

[2] Some industrial turbines use water or steam injection to aid in emission control.
[3] Some aviation turbines deposit dense pyrolitic carbon at the combustor, and eventually this detaches and passes through the turbine where it may cause erosion or foreign object damage.

because of the short residence time and in spite of the overall excess of oxygen. If the fuel contains sulfur, which is the case for most hydrocarbon fuels or sulfur-containing natural gases, this will normally oxidize during combustion to SO_2, which may then oxidize further to SO^3 as the temperature drops. The product mixture is usually described as SO_X. Finally, and most importantly, oxides of nitrogen may be formed during combustion. This can result from the oxidation of the nitrogen in the combustion air or by the oxidation of fuel-bound nitrogen in liquid fuels. Again, different oxides may be formed, and the mixture is described as NO_X.

The formation of NO_X is favored by high temperatures in the combustion zone and by a significant excess of oxygen. If there is less than a stoichiometric amount of oxygen in the primary combustion zone, NO_X levels will be low; but CO may be present. As a result, there is often an inverse relationship between NO_X and CO (and unburnt hydrocarbons) in the exhaust gas. For land-based gas turbines, local environmental legislation may limit the amount of these contaminants that can be emitted and, as a result, low-sulfur fuels are generally specified. This is true for gas-burning engines whether the gas is natural gas or coal-derived gas.

For coal-derived gas, the fuel-bound nitrogen is converted to ammonia, which is also generally removed from the gas during the clean-up process. Coal also contains some alkali metals, sodium and potassium, and the degree to which these are removed depends on the clean-up processes.

Control of NO_X can be achieved by reducing the maximum flame temperature, for example with water or steam injection. This does not reduce the turbine inlet temperature. The increased mass of hot gas will result in an increase in specific power output but lower thermal efficiency. The NO_X emissions can also be controlled by a catalytic reduction process for the exhaust gas, although this does represent a significant penalty in cost and efficiency.

An innovative concept for advanced transport engines is a "rich-lean burn" concept. The modern "dry, low-NO_X" burners often used in current generation land-based gas turbines use some version of a two-zone combustor. The first, or primary, zone has a stoichiometric diffusion flame that stabilizes the combustion process. The second zone contains a lean fuel-to-air mixture that burns at temperatures too low to produce NO_X. A small amount of NO_X is still produced in the primary zone.

Coatings such as TBCs lower emissions by reducing the need for cooling air.[4] The less air required to cool the combustionliner, turbine vanes, and turbine blades, the more air available to dilute the fuel-to-air ratio and achieve lean combustion. For example, the wall temperature of a typical can combustor is moderated by cool air flowing over the inner surface, which is known as "film cooling." Film cooling can quench gaseous reactions in the combustion process near the wall and lead to enhanced CO and unburnt hydrocarbons emissions. TBCs [5] can be used to shield the combustor liner from the combustion heat, reducing the amount of cooling air needed and lowering the amount of CO and unburnt hydrocarbons formed. This effect can be further enhanced by the use of a double-walled combustor, where the inner wall is solid and protected by an inner TBC, and the outer wall is perforated to allow impingement cooling of the inner wall.

CONCURRENT COATING DEVELOPMENT

To meet the very demanding and constantly changing requirements for high-temperature structure to support the next-generation engines, revolutionary changes in hot-section materials and coatings will be necessary. Choosing a high-temperature coating and determining the next step in research and development rank as a complex, iterative process. For example, engineers must consider cost-benefit scenarios as well as structural factors in evaluating design choices. These scenarios must include repair and maintenance costs-as well as off-line user costs-within the context of costly, idle assets and contingency plans in the event that downtime exceeds the amount allotted for customer services (e.g., electric power or scheduled airline service). If each step of the iteration were performed sequentially, significant time would elapse before a design could be completed.

As a result, the sequential, separate development of structural alloys and coatings is no longer adequate. The sequential method is being replaced by concurrent development of the total system since overall system performance is of primary importance to the users of turbines. Thus, the design and selection of structural materials and their coating systems should be performed concurrently by teams of experts drawn from each critical function (e.g., design disciplines, materials engineering, manufacturing, and cost avoidance). Integrated process development, or concurrent engineering, is becoming an imperative for coatings and will remain an imperative in the future. Life-prediction methodologies that are based on a fundamental understanding of the key parameters that arise from engine design, anticipated application use, coating and superalloy properties, and prior engine experience databases, could also be useful in the design of future engines. The needed information must be available, properly measured, and defined.

---

[4] It was thought at one time that TBCs reduced emissions directly, but there is little evidence to support this view.
[5] TBCs have been used on combustion liners for many years to provide protection from hot streaking.

6

Refurbishment of Coated Structure

As described in chapter 4, a coating degrades at two interfaces: the coating/gas path and the coating/substrate. Deterioration at the coating/gas-path interface is a consequence of oxidation or hot-corrosion mechanisms that occur at temperatures well below the incipient melting point of conventional superalloys. As the temperature of the surface metal increases, solid-state diffusion at the coating/substrate interface causes compositional changes that can compromise coating protection and substrate microstructure, resulting in markedly reduced component life. Coatings are usually designed to wear out without causing degradation of the underlying component. The conventional approach has been to ensure that the coated component remains functional until overhaul, at which time it is fully assessed. For airfoils operating at temperatures below the level at which diffusion is a concern, recoating with or without rework allows the part to be returned to service. Hence, the coating serves as a renewable surface that can extend component life.

The goal of component refurbishment is to restore component integrity in an economical and timely fashion. To accomplish this goal, engineers rely on visual nondestructive evaluation (NDE) techniques to assess the component condition and develop a schedule of required repairs. Safety, reliability, and economic factors govern the type and extent of permissible repairs. Repairs therefore vary greatly depending on the type and criticality of the component to be refurbished (Haafkens, 1982). Replacement of the coating is generally a small portion of the overall repair but plays a key role in ensuring that the component endures for the remainder of its expected service life. The repair of aircraft turbines differs significantly from that of land-based turbines. Repair facilities and practices for aircraft engine components must be certified by the military or the Federal Aviation Administration. This generally results in industrywide minimum standards for the quality of the repair, the repair facility, and the personnel performing the repair. No such standards currently exist for repair of industrial gas turbines.

Because of the use of thin walls and the highest-strength superalloys, aircraft turbine blades have had limited tolerance in the past for high-temperature oxidation and virtually no tolerance for hot corrosion once the coating has been breached. For this reason, highly distressed aircraft turbine blades are not repaired. Industrial gas turbines, which employ thick walls and lower-strength alloys with a higher resistance to corrosion, have a higher resistance to oxidation and hot corrosion once the coating has been breached. In this case, the blades may be able to be refurbished and returned to service if the component is not severely distressed. The increased complexity of blades in the new generation of engines will make this more difficult.

FACTORS AFFECTING COMPONENT LIFE

For all gas-turbine applications, the minimum total service life as well as the life until an inspection interval must be determined by engineering analyses, component testing, and any justifiable combination of verifiable methods. Determining life expectancy often requires complex analysis of independent and interacting factors relating not only to operating conditions but also to the reliability of the system and the quality of the finished components.

When should a component on a gas turbine be repaired or replaced, and when should the component continue to run? These critical run/repair/replace decisions are based on a combination of turbine manufacturer recommendations, prior experience, and, more recently, detailed analysis using modeling tools. These decisions are influenced by the amount of risk the user is willing to tolerate,[1] the need for the equipment to continue operating, and the maintenance budget.

Run/repair/replace decisions are not an exact science, and a great deal of judgment can be involved. There has been recent success with expert system software that assist the equipment operator and others in making these decisions. For instance, the Electric Power Research Institute has developed software based on engineering analysis along with calibration and verification of the predictions by the examination of field-run hardware (Bernstein, 1990).

The role of coatings in run/repair/replace decisions is crucial since the condition of the coatings often determines the service life of the component. Thus, assessing the coating condition by calculation or inspection is critical. Algorithms for coating degradation are under development for some of

---

[1] Recommendations from turbine manufacturers and others are sometimes interpreted by engine operators in terms of the perceived level of their experience and the economic benefits from additional sales of parts or services.

the industrial gas-turbine engines. Appendix C, Survey of Nondestructive Evaluation Methods, and appendix D, Modeling of Coating Degradation, contain further information that is summarized in this chapter.

REPAIR OF HIGH-TEMPERATURE COATINGS

This section addresses the repair of coatings for industrial turbine hot structures. The criteria and practices are similar for aircraft engine components, except that aircraft engines traditionally have made more extensive use of advanced alloys, incorporated a higher degree of design sophistication, and had greater regulatory oversight requirements because of flight safety concerns. These factors generally have resulted in restricting allowable repair options for aircraft engine components.

Importance and Need of Repair Technology

The single largest expense in maintaining industrial gas turbines is the cost of new hot-section components, particularly the blades and vanes. Consequently, all possible efforts are made to repair these components instead of replacing them. Coating replacement is one of the necessary steps in the repair of these components. Coatings must typically be removed and reapplied a minimum of once during service life and, preferably, several times. Internal coatings must be capable of surviving the repair process intact or being renewed.[2]

It is unclear if a new coating must be capable of being refurbished in order to be a viable, commercial coating, however. The unit cost of generating energy is the sum of the equipment acquisition and siting costs, fuel costs, and maintenance costs divided by the energy produced. Maintenance costs include engine teardown and assembly, inspection and repairs, new parts, and the cost associated with the unavailability of the engine. If a new coating could allow higher-temperature operation (for increased efficiency), the savings in fuel costs could possibly outweigh the added expense of purchasing new parts versus repairing old parts. Consequently, the ability to repair a coating may not always be a primary consideration in developing a new coating

Current Status

Although details of repair procedures may vary, the repair steps themselves are very similar. Acid baths usually remove aluminide coatings, while manual belting (sanding) or electrochemical processes remove overlay coatings. Water jets have also been increasingly used to remove both aluminide and overlay coatings. The main disadvantage of the acid baths is that they contain toxic chemicals that must be carefully handled and disposed; manual belting is costly, labor-intensive, and difficult to control precisely. The repair procedure must remove all of the coating and deteriorated base metal; incomplete removal can seriously shorten the service life of the new coating.[3] Heat-tinting procedures work reasonably well for detecting remnants of the old coating. Advanced, automated NDE inspection methods are needed to verify that a clean substrate has been prepared for recoating operations.

The recoating operations generally follow the same procedures and technology as the application of the original coating. The durability of the reapplied coating is not well quantified. Coatings applied over the base metal should be expected to have the same durability as the original coating. Diffusion coatings applied over weld and braze repairs may have lower durability, since the composition of the weld or braze are different from the base metal. Overlay coatings, which depend less on the base metal for their final composition, are the least sensitive to these effects. However, manufacturing processing sequences, such as grinding blade root platforms after coating, can present further challenges for recoating operations.

Future Directions

The most important need for the repair of industrial gas turbine components is industrywide repair specifications and regulations of the quality of the repairs, developed by an independent, knowledgeable, and unbiased organization. These specifications and regulations should include the removal and reapplication of the coating. Improved NDE is needed to determine when the coating has been removed and when there is no base-metal attack.

The durability and properties of refurbished coatings are not known but are of great importance to gas-turbine operators. Changes in airfoil composition and structure during repair cycling (e.g., brazing or welding) can affect subsequent coating systems. As part of the knowledge, engineers need to assess the durability of coatings applied over weld and braze repairs. Methods are needed to make local repairs of coatings during both manufacture and operation. During manufacture, some areas of the coating may fail specification or become damaged. Methods to inspect these areas are needed. Some areas of the coating may also become damaged during operation. If local repair procedures are available, these areas could

---

[2] Service run coatings are typically not locally repaired, although examples of these "mini-repairs" do exist.
[3] It can be difficult to detect base-metal attack that is confined to the grain-boundary region.

be repaired on site and the component (e.g., turbine blade) returned to service.

The future of aircraft engine repairs will most likely parallel those of the industrial turbines-with the added complexity of thinner walls and more sophisticated cooling passages (e.g., cast cool/multiwall quasi-transpiration cooling concepts). Thin walls in advanced components can preclude any stripping of the old coating, making it unrepairable.[4] Development of advanced NDE techniques will be essential for assessing components that are expected to be multiwall/thin-wall structures with multilayered coatings, and the coating will be an integral part of the component design and manufacture.

The ability to repair TBCs, increasingly used to improve performance, has yet to be fully assessed. Removal and reapplication of the bondcoat should follow the same steps as for any overlay coating.

STANDARD DESIGNATIONS FOR COATINGS

Although generic types of coatings exist (e.g., aluminides, chromides, and TBCs), engineers have no standard system of designating or defining coatings. Instead, each manufacturer and vendor has their own commercial nomenclature, meaning the same coating may have ten or more different names. This state of affairs causes confusion for gas-turbine owners, who have to select coatings to refurbish their components. To appreciate this confusion, one only has to imagine the situation that would exist if there were no accepted standard designation system for steels.

For example, the U.S. Air Force believes that common designations for coatings are needed to assist in the procurement of repairs. In the industrial gas-turbine community, the large number of coating choices is overwhelming those in charge of maintaining the engines. A system for designating coatings would significantly reduce the current state of confusion. Other benefits of a system that designates coatings include increased competition in the marketplace, simpler specification of repairs to gas-turbine components, and increased access to foreign markets.

Developing a designation system for coatings could follow a path similar to that used for metals. The American Iron and Steel Institute (AISI) system for steels provides the nominal chemistry for each type of steel and is used universally. The Aluminum Association system for designation of aluminum alloys is similar to that of steels but includes suffixes for heat treatments and other processing. This system for aluminum alloys is also used universally. A different approach is taken by the American Society for Testing and Materials (ASTM). This organization develops consensus standards for materials and other items. Its material standards, which simply use the standard number and often a subcode, designate the chemistry, product properties, and sometimes the processing. While not intended to be a standard designation system, the ASTM standard number has achieved the status of a material designator, such as A36 steel. Existing coatings standards published independently by a number of organizations are listed in appendix A.

Compared with bulk metals, a coating has some unique aspects: it depends on the substrate and the process, it can be a composite of two different coatings, and it evolves relatively rapidly. There are enough similar coatings in common use that a designation system is practical and beneficial. Any system must have the capability of adding new coatings as they are developed. The three systems described above for bulk metals (i.e., AISI, Aluminum Association, and ASTM) have this capability. Some examples of potential designation systems for coatings are summarized in appendix F.

NONDESTRUCTIVE EVALUATION

NDE aims to identify defects, qualitatively or quantitatively, that could lead to failure, while not changing the material being tested. It is often desirable to make these measurements without direct contact with the component. Recently, this goal has been extended to include:

- assessing the after-fabrication condition of the coating system
- aiding the characterization of advanced materials
- developing process control methods to minimize the variation in quality of a manufactured part
- ensuring safe operation by assessing the in-service condition of coated components and remaining life
- evaluating the components after repair

All NDE methods use some external source to produce a measurable response from the sample without causing a permanent change in the specimen. The information extracted by the measurement is determined both by the initial source/specimen interaction and the detection method. The combination of source and detection methods comprises an NDE technique. Table 6-1 lists some of the primary NDE methods used for coating inspection. Appendix C provides more detailed discussions of these techniques.

The focus for the application of NDE methods should follow the five areas identified above. Constituencies in each area have an interest in NDE methods; hence prioritization is difficult. At the least, it is important to retain the traditional NDE focus on initial qualification and in-service inspection while examining opportunities in the other areas. The priority should be on methods that can address the key issue of

---

[4] Some of the current turboprop/turboshaft high-pressure turbine blades are not repairable.

TABLE 6-1 Survey of Nondestructive Evaluation Techniques

Source Technique	Energy Form (parameters and limits)	Selected Uses
Visible Light	Reflected/transmitted light	Surface characterization
Visual inspection	Low spatial resolution semiquantitative	Surface damage of substrate or coating
High resolution imaging (microscopy)	Spatial resolution of surface features =1 µm	Metallography, grains, surface breaking cracks
Spectroscopy	Wavelength variations in reflection/transmission	Contaminants, oxide layers, composition
Ellipsometry	Polarization state of reflected light	Thickness of surface films dielectric properties
Infrared	Reflected/transmitted light	Surface/bulk characterization
Imaging spectroscopy	Spatial variations of reflection/transmission wavelength dependence of optical properties	Spallation of TBC; structure of metallics; composition of coating; impurity type
Emission spectroscopy	Wavelength dependence of emission from heated materials	Emittance versus wavelength for TBC and other ceramics
X-Ray Transmitted Energy	Attenuation/diffraction in bulk	
Radiography	Integrated attenuation along path	Voids; thickness changes in metal wall
Diffraction	Unit cell parameters, long-range lattice order	Plastic deformation; residual strain
Topography	Single lattice reflection, spatial variations of lattice order	Spatial estimate of crystallinity of single-crystal blades/vanes
Microwave		
Imaging	Spatial changes in reflectivity of coating and substrate	TBC contaminants, coating thickness, substrate conductivity
Spectroscopy	Wavelength dependence of optical properties	TBC coating thickness, rotational relaxation times
Electron Source	Characteristic X-rays	Composition thickness
Backscatter	Energy selective detection of fluorescing X-ray photons	Elemental composition, some lateral resolution, layer thickness
Electromagnetic Eddy current	Magnetic fields from induced/injected currents	Factors affecting conductivity and permeability
Photoinductive imaging	TBC coating thickness; lateral resolution = 250 µm	Thickness of substrates and coatings
Magneto-optic imaging	Lateral resolution = 10 µm; spatial variations in magnetic field of induced currents as modified by cracks, holes, etc.	Substrate crack detection residual stress; work hardening; coating thickness; crack location/sizing; visualization of conductivity-related defect regions in conducting materials
Thermal with Particle or Electromagnetic Source	Bulk and surface properties affected by temperature	Thermal parameter disbonds, voids, cracks, thickness
Imaging	Spatial variations temperature < 10 mK, emissivity > I µm	Coating thickness; coating adhesion; thermal properties of coating and substrate
Spectroscopy	b(l), R(1), e(l) throughout the electromagnetic spectrum	Composition of substrate and coating transmission regions emissivity

interface degradation, preferably without touching the specimen. Although some researchers have developed methods that might be used to assess interfaces, studies for the most part have been preliminary and rather uncoordinated. Efforts should be better focused and carried out at a level where some of the newer methods could be brought into practice.

7

Near-Term Trends and Opportunities

Coating technology for high-temperature turbine engine structure is advancing at a rapid rate. Engine design engineers, materials scientists and engineers, and coating technologists are actively seeking the best alternatives to provide improved engine performance at the lowest possible cost and risk.

To support the development of advanced aeronautical, industrial, and marine turbines, coating and substrate materials are being increasingly developed concurrently. Until recently, the evolution of high-temperature coatings had been largely independent of component development (e.g., alloy composition, casting process, and cooling design). This sequential development method will be inadequate for advanced coating systems and systems where the coating and the substrate are becoming an integral entity from a life-cycle perspective (e.g., design, materials and process development, manufacturing, and product support). For example, concurrent development is helping match the current generation of MCrAlY coatings to new nickel-base superalloys, as well as developing TBCs (thermal barrier coatings) to be used with a MCrAlY undercoat and superalloy substrate.

This chapter summarizes current trends in coatings development, focusing on coatings expected to be incorporated in engines within the next five to ten years. It also identifies critical areas of research and development that could rectify shortcomings in coating technology. The underlying theme of this chapter is that advances in the areas highlighted in this chapter provide the knowledge for realizing concurrent engineering of coatings and substrates.

THERMAL BARRIER COATING DEVELOPMENT

TBCs are finding increased application in overall component design. Over the past 25 years, cooling technology has increased turbine operating temperatures by roughly 400°C (720°F; Soechting, 1995). Superalloy material and processing advances have increased operating temperatures by approximately 120°C (215°F) because of the progression from equiaxed superalloys to third-generation, single-crystal superalloys. Further advances will be possible with these materials through the use of even more sophisticated cooling and casting schemes (Nealy and Reider, 1979; Turner et al., 1992; Caccavale and Sikkenga, 1994), in conjunction with widespread use of TBCs.

Using insulating TBCs, nickel-base superalloys have been shown to support a metal temperature reduction of as much as approximately 140°C with a 5-mil-thick TBC (Manning-Meier and Gupta, 1992). Thus, these coatings have the potential to extend the use of superalloys into advanced engine applications. In the near term, the committee believes that TBC technology merits most of the development and application effort. This will allow the current generation of high-temperature structural materials to bridge the requirements gap until substrates capable of functioning at higher temperatures are fully developed. However, the insulative effect of TBCs can be dependent on the substrate (e.g., because of substrate thermal conductivity). As with any coating/substrate system, the coating must be tailored to the substrate material and component design. *Also, to take advantage of the reduction in substrate-metal temperature provided by TBCs as a way to optimize turbine efficiency, TBC reliability should be considered a critical factor in need of further development.*

Generating a more compatible and oxidation-resistant bond between the metallic substrate and the TBC will also need continued effort. Pratt & Whitney Aircraft (Novak, 1994) has demonstrated improved compatibility in turbine shrouds. In this application, a coating consisting of a graded layer acts to increase strain tolerance. Although this approach has been shown to be effective where thick (greater than 0.050 inches) coatings are tolerable, application to rotating components, such as airfoils, is usually impractical because of the high inertial mass added by the coating (see chapter 5). Furthermore, for use at higher temperatures, the effects of graded layer oxidation will need to be negated in order to make graded coatings feasible for future use in aircraft engines. *The development of durable bondcoats that can accommodate thermal expansion mismatch, combined with improved interface oxidation resistance, should be the focus of near-term research.*

Current knowledge of TBC failure mechanisms lies primarily in empirical data. Although the NASA HOST reports show significant progress toward defining a semi-empirical life-prediction methodology for TBCs, surprisingly little is understood about TBCs in the critical areas of radiative and conductive heat transfer and emissivity, failure mechanisms, and fundamental physical and mechanical properties (i.e., fatigue, monotonic properties, and time-dependent properties).

Research is needed to improve the understanding of TBC behavior (and thereby improve TBC performance) as well as to provide reliable information for quantitative modeling.

Until recently, the energy transport process in ceramic TBCs has been measured by an empirical parameter: thermal conductivity. This measure does not distinguish the relative contributions of radiation and true conduction. To determine the relative importance of radiative and conductive transport in ceramic TBCs, a number of factors affecting the two transport processes must be considered (see appendix B). Attention should be given to understanding the mechanisms of energy transport in TBCs with the goal of reducing energy transport rates and increasing substrate-metal protection even further. Knowledge of the relative importance of radiation and conduction will guide research strategies aimed at improved coating performance. *Attention should be given to understanding the mechanisms of energy transport in TBCs with the goal of reducing energy transport rates and increasing the life of components.*

The performance of TBCs depends on the characteristics of interfaces with well-understood and controlled features. For example, the thermally grown oxide interfacial layer between the metallic bondcoat and the ceramic TBC influences adhesion and hence coating life. *Improved understanding of interfacial behavior is required to control properties and predict performance. A more compatible and oxidation-resistant bond between the metallic substrate and the TBC should receive continued near-term emphasis.*

COATING PROCESSES

The current generation of MCrAlY coatings, as well as those emerging from new TBC technologies, would benefit significantly from advances in process control. Both types of coatings are deposited by the same processes and both show a similar variability in performance. *Improved on-line control should be developed to ensure that the coated structure is highly reproducible, a necessity if structures are to perform within the parameters set by service requirements. For TBCs, it is not clear what processing factors are most important to performance or performance variability. Consequently, these must be identified before appropriate sensors can be incorporated.*

To manufacture improved coatings, process modeling, monitoring, and feedback controls will all have to become more common. The near-term TBC effort should focus on two fronts: minimizing variability while maximizing service life. The technical approaches by which developers choose to tackle this two-pronged challenge will no doubt vary. Iterations of models followed by increased understanding of parametric variables, and then model modification, will be necessary. Manufactured product uniformity will, in addition, require the development of process sensors and feedback controls, again perhaps, by an iterative process. Application of laboratory NDE (nondestructive evaluation) concepts, such as infrared imaging, should be applied to manufacturing process control (Murphy et al., 1993).

The two principal manufacturing processes for TBCs (i.e., plasma spray and physical vapor deposition) produce coatings with significantly different microstructures, properties, and durabilities. In addition to these technical attributes, there are significant coatability issues that can strongly impact the cost of the coating. *The choice of coating process depends on a balanced assessment of the technical attributes, the coatability issues, the cost, and the performance requirements demanded from the coated component.*

Process control can only be achieved if the relationship of the process-to-product performance is understood. Development of such knowledge should be a priority, near-term endeavor for advanced coatings. Such parametric data, guided by an understanding of coating behavior and failure modes, can be quickly and reliably acquired by rig testing, but the variability of the rig test must also be known. Rig-test modeling, parameter sensing, and feedback control will be an evolutionary and continuing near-term activity.[1]

Although engine tests do not necessarily provide data on individual processes, they do present opportunities to test complete (integrated) concepts in both propulsion and stationary power plants. These tests are necessary to provide an overall qualification of a new coating system.

Repair and Overhaul

Engineers rely strongly on visual inspection to assess the condition of coatings. *Improved NDE is needed to determine when coatings should be removed and the extent of base-metal attack* For aircraft engines, the development of advanced NDE techniques and cost-benefit models will also be essential for assessing the new generation of components. These components are expected to be multiwall/thin-wall structures with multilayered coatings used as an integral part of the component design and manufacture. NDE, or some form of monitoring, is required to assess coating health during use. An example of a simple monitoring method for static structures would be to use sensors to detect abrupt increases in temperature, indicating potential TBC spalling and a need for inspection. Other more complex schemes may be available.

Analytical Methods

Because future engines will rely heavily on coatings to protect hot-section components, accurate models for coating

---

[1] The NRC (1989) study On-Line Control of Metal Processing predicted this process evolution.

and component life will become essential. Coating life and inherent substrate environmental resistance are key determinants in setting the frequency of engine inspections and overhauls. Coating producers and users thus have a widespread need for general purpose data and models. Examples of these include data on long-term thermodynamic and structural stability, generic process models, and life-prediction models. *Engineers need advanced analytical techniques to measure coating properties, including those for nanostructure materials.* Appropriate combinations of methods must allow measurement on the microstructural scale of the material relevant to specific properties (e.g., adhesion, fracture toughness, thermal conductivity, and elastic modulus). Such methods may provide the basis of standard tests for assessing coatings.

Analytical design analysis of the mechanical stress (strains) imposed on coated components is fairly well understood and may help in estimating service life. The degradation caused by oxidation and, to a lesser extent, corrosion is reasonably well defined but not with high precision. The impact of other factors, such as erosion and variable environmental conditions (e.g., variability in air quality), is less clear. For example, the prediction of special events, such as impact damage, are not currently factored into life models. Both users and manufacturers have a strong interest in life prediction, and refinement of these methodologies is a continuing task. The precision of these models should improve with experience. Indeed, a better understanding of current coatings and substrates will probably be gained more from service experience than from additional laboratory work.

Failure to properly predict service life from laboratory tests has often resulted from variability in rig test conditions. While rig conditions may vary, such variations are probably less than those occurring in service. Service-life variations may also result from a factor far less understood: manufacturing variability.

There is a broad-based need for prestandards research that, particularly for TBCs, identifies critical properties and the scale on which they are relevant. This research provides the basis for formulating a set of standards used to measure and compare coating systems. It can also provide the data required for performance models (e.g., methodologies for measuring thermal conductivity, interfacial adhesion, and microstructural characterization).

Modeling

Because future engines will rely heavily on coatings to protect hot-section components, accurate models will become essential to describe a number of coatings-related requirements, particularly (1) process attributes, (2) degradation modes (e.g., oxidation and corrosion), and (3) service-life prediction and residual-life assessment. Process modeling is most important to coating manufacture and should be able to characterize process-dependent factors such as microstructure, rate of coating deposition, and cost, all based on user needs. Degradation modeling is most important to coating design and development. Coating life and inherent substrate environmental resistance are key determinants in setting engine inspection and overhaul intervals. Few models in the public domain address any of these needs.

Coating life is determined by many factors: the mission profile, environmental conditions, accuracy of the design analysis, the coating system selected, and the manufacturing process used. At this time, there is a good-to-excellent understanding of the design conditions and mission profiles and a fair-to-good understanding of the environmental conditions for conventional applications. Understanding the relationship between process parameters and in-service performance (scatter of the properties) is fair to good for metallic coatings and currently only poor to fair for TBCs. Refinements in life prediction will be gained through models of the entire coating life cycle, ranging from engineering analysis to component retirement. The key factors needed to complete this model are the integration of current, specific domain models, better comprehension of special mission-caused (random) events, and a parametric understanding of the manufacturing process. The greatest benefit will probably be gained by defining manufacturing process capability and control, along with an improved understanding of environmental conditions and their effects.

Nondestructive Evaluation

Advanced NDE methods are needed to characterize hot-section turbine components at inspection points during their service life. Some methods are currently available to support decisions on repair or replacement of components, and their application should be encouraged. In other cases, methods need to be developed for this use and as a basis for life prediction.

How useful are advanced NDE methods in assessing the fitness of turbine components following repair and refurbishment? A comprehensive review of the currently used and candidate NDE methods could be applied to in situ process monitoring and control, enhancing the reproducibility and cost of high-temperature coatings.

8

Long-Term Opportunities and Innovative Systems

In the long term (i.e., beyond five to ten years), the static and dynamic components of high-temperature turbine engines will continue to operate in harsh environments at temperatures above the incipient melting point of nickel-base superalloys. Two primary factors are likely to increase research and development activity in advanced hot-section materials. First, the realization is spreading that the current material systems for coated substrates are technologically mature (see chapter 7). Second, market demand is growing for turbines with even higher efficiencies, in the aircraft engine and land-based applications, to attract the necessary investment. The required performance levels will only be attained if one or more of the following occurs:

- use of advanced substrate materials (e.g., ceramics)
- production of cooling schemes that impart a very low performance penalty
- development of highly reliable TBCs (thermal barrier coatings) or some other insulating technologies

Monolithic-ceramic and ceramic-composite substrate materials represent a long-term path to enhanced high-temperature performance. These materials, however, will likely need protective coatings for long-term durability in some operating environments. For example, silicon carbide and silicon nitride are susceptible to rapid high-temperature oxidation (as discussed in chapter 3).

This chapter presents several innovative coating concepts that are attractive but unproven. Many are ideas borrowed from other technology areas, such as microelectronics. They are offered to stimulate further innovative approaches by researchers in industrial, university, and government laboratories. The committee believes that such imaginative ideas will lead to the necessary breakthroughs for the future. The first section of the chapter presents a few architectural concepts for advanced coatings. The second section discusses several innovative coating technology concepts.

INNOVATIVE COATING ARCHITECTURES

As stated throughout this report, close integration of coating and substrate materials will undoubtedly occur. The coated structures emerging from this integration will be known as *hybrid components,* since the demarcation between substrates and coatings will be difficult to discern. Control of interface properties for these new coatings will be a critical feature, especially the control of the substrate/coating junction. The control of this singular interface will continue to be a key to the success of high-temperature coating application. Advanced coatings will have multiple interfaces that must operate at higher temperatures in steeper thermal gradients and will have to be confined to the sub-micron scale. Such structures are referred to as nanostructures, with layer thicknesses in the 1- to 1,000-nanometer range. Adherence and metallurgical stability will also be major requirements. In all cases, an important element is the stability of such structures in the high-temperature environment of an operating turbine engine. *Research on issues of stability is a high priority.* The structural motifs for hybrid coatings include continuously graded coatings, horizontally layered materials, interphase layer materials, vertically layered materials, and three-dimensional structured materials.

Continuously Graded Coatings

The area of graded coatings (e.g., layered, continuously graded, and micro-or nanolayered coatings) offer the potential to improve coating performance by providing position-dependent physical or mechanical properties. The concept of graded coatings-more recently termed functionally graded materials (FGM; Ramesh and Markworth, 1993)-was developed and used to improve TBC adherence in rocket engines over 30 years ago (Ingham and Shepard, 1965). The same types of thermal spray coatings, which included both layered and continuously graded thermal spray coatings, have since been used as outer air-seals of aircraft gas-turbine engines and thick-graded TBCs in diesel engines (Miller, 1990; Goward, 1987; Yonushonis et al., 1987). While current thick-graded coatings are inappropriate for rotating parts in turbines, the continued improvement in process control in thermal spray and other deposition methods (appendix E) offers the opportunity to develop thin-graded coatings that may offer some of the benefits of grading while being sufficiently light for use on turbine blades.

FGM coatings will minimize thermal stresses in the coating caused by thermal coefficient mismatches. A continuous compositional variation across the coating is associated with a continuous variation in the coefficient of thermal expansion (CTE). In the ideal case, the composition of the coating would continuously vary, becoming a pure ceramic at the exposed face of the coating. The CTE of the coating region adjacent to the substrate would match the expansion coefficient of the substrate.

The inherent inhomogeneity in the microscopic scale of these materials raises questions about their stability at high temperatures. These questions must be answered to establish the viability of these approaches. *To spur development in this research area, there should be a critical assessment of the use of FGMs as coatings and a definition of the influence of multilayer and nanostructure morphology on resulting properties.*

Hand-in-hand with advanced substrate materials (e.g., ceramic carbide and nitride, intermetallics, composites, and refractory metal alloys), there exists a need for coatings capable of withstanding high temperatures.[1] Meeting this demand may require alternative coating materials as well as alternative application processes. Advanced substrates will possess coatability characteristics (e.g., diffusion rates and surface chemistry) dramatically different from current superalloys. *Novel concepts may be needed for coatings and surface treatments to protect future advanced substrate systems.*

Horizontally Layered Materials

Layered materials offer a number of opportunities for advanced coating concepts. They can be designed to reflect thermal radiation, reduce heat transfer across an interface, and match the CTE of the coating with the substrate.

A coating may serve as a high-temperature, optical multilayer interference filter that reflects radiation. Such a system may consist, for example, of alternating thin layers of two oxides, each having a different refractive index (e.g., zirconia-yttria 2.07 and alumina 1.60). A process similar to Bragg diffraction reflects a very high fraction (e.g., 0.95) of the incident radiant energy. The range of thicknesses of the deposited layers is related to a quarter wavelength of the peak wavelength of light (about 1 to 2 microns), which corresponds to the black-body temperature of the engine. Calculations show that 95 percent of radiant energy will get reflected for only 38 microns of total coating thickness. This corresponds to about 63 pairs of oxide layers.

Multilayered coatings may also reduce the heat-conduction load to the substrate by taking advantage of the lower rate of heat transfer across an interface. Heat transfer across an interface is reduced if the two layers have differing thermal effusivities.[2] In a multilayer structure, the successive reductions at a series of interfaces may lead to a significant reduction in the total heat transfer by conduction.

Multilayers also provide the opportunity to match CTE at the coating/substrate interface, possibly minimizing the thermally induced internal stresses in the coating. The CTE of the coating layer adjacent to the substrate would closely match the expansion coefficient of the substrate. In subsequent layers, the composition would vary in step increments, becoming a pure ceramic at the exposed face of the coating. This concept will need to be examined from an engineering mechanics point of view, however.

Interphase Layers

The boundary between the substrate and the first layer of oxide is the crucial interface in any system designed to afford protection in a high-temperature environment. Separation of the oxide from its substrate is less likely to occur if this interface is replaced by an interphase, the composition of which could vary smoothly and gradually from the substrate metal (or ceramic) to the full oxide of one of the abundant elements of the substrate.

The full oxide layer immediately adjacent to the substrate is currently alumina (often with some chromia). In present technology, zirconia, usually stabilized with yttria, is then deposited to provide a thermal barrier. From the standpoint of thermal expansion, zirconia might possibly adhere better if the MCrAlY/zirconia or alumina/zirconia interface were compositionally graded through intermediate compositions. However, previous research has shown that such a compositionally graded metal/ceramic layer undergoes oxidative expansion, resulting in severe compressive stresses that cause the oxidized graded layer to buckle the coating from the substrate (Duvall and Ruckle, 1982; DeMasi-Marcin et al., 1989). There may be potential for other gradient schemes, however. Zirconia is softer and susceptible to erosion; it might be worthwhile to *top* the zirconia with a graded layer as well, although this concept must be balanced against weight considerations.

Vertically Layered Materials

Using a layered structure in the coating or substrate also improves the thermal isolation of the substrate from the engine environment. The present design for TBCs appears to have conflicting demands—the need for adherence and for

---

[1] Higher-temperature-capable ceramic coatings are currently being explored, for example, as part of the High-Speed Civil Transport program.
[2] Thermal effusivity is defined as the square root of the product of thermal conductivity and specific heat.

low heat transfer between coating and substrate. Adherence to the substrate (required for structural integrity) invariably results in good transfer of heat to the substrate. One potential solution is to take advantage of the lithographic patterning technology widely used in the semiconductor industry.

For example, the substrate could first be coated with an adherent ceramic material that has a high thermal conductivity but can withstand high-temperature oxidation (Step A). Step B would use existing lithographic patterning techniques to form high-aspect-ratio (high depth with narrow width) trenches in the coating. The following step (Step C) would fill in the trenches with a poor adherent and thermally low-conducting material. This would limit heat transport down to the substrate. The final step (Step D) would seal the outer surface.

Such a channel structure may also reduce the radiative heat load. If the repetition distance between channels is commensurate with the infrared wavelength for which the unstructured coating is transparent, diffraction from the channel structure will increase the reflectance of the coating/substrate system regardless of the intrinsic reflectivity of the substrate.

Three-Dimensional Structured Materials

Employing three-dimensional networks to coat microstructure can improve the mechanical stability of the coating and possibly its resistance to heat conduction. An example is the distribution of ceramic whiskers in a ceramic matrix of a similar composition. The whiskers could be formed in situ through a phase transformation process (e.g., silicon carbide [SiC] and silicon nitride [Si_3N_4] precipitate in an acicular shape [needles or whiskers] when they transform from their beta form during heating).

In situ formation of three-dimensional ceramic reinforcements may be applicable to coatings of ceramic-matrix composites. SiC, Si_3N_4, or other ceramic materials might be incorporated as aggregates or agglomerates of beta-form particles in the matrix powder and blended with the matrix phase before heating.[3] On heating, the aggregate phase will transform from the beta to the alpha form and precipitate as a whisker or needle. In this way, sintering, densification, and the desired amount of bonding between the matrix and whisker phases can be achieved, yielding a robust composite coating that is strong, lightweight, highly dense, and resistant to severe environmental distress.

OTHER INNOVATIVE CONCEPTS

The committee identified several additional innovative technologies that may improve the coating/substrate system. In most cases, researchers have examined these concepts, but the work has not advanced to the point where practical systems have been developed. The committee believes that further research is warranted in these endeavors despite the lack of success.

Nanocrystalline Coatings and Substrates

Promising research of nanocrystalline substrate and coating materials, with improved high-temperature properties, may allow for higher engine-operating temperatures and therefore improved performance in the future. For example, the SiC nanoparticles in SiC-particle-reinforced alumina appear to facilitate crack healing, resulting in improved high-temperature strength and creep resistance as compared to monolithic ceramics (American Ceramic Society Bulletin, 1996). Nanocrystalline coatings consisting of TiN nanocrystallites embedded in amorphous Si_3N_4 are being studied for use as a wear-resistant coating (Dias et al., 1995). Nanocrystalline structures could also potentially be developed to provide high reflectivity to reflect radiant heat or lower conductivity because of phonon scattering.

One potential problem with using nanocrystalline materials at elevated temperatures is exaggerated grain growth, which could result in sintering to great density (Dowding and Malghan, 1995). Sintering is problematic for TBC coatings in particular, because a relatively porous microstructure is typically desired to accommodate strain release. Fine grain size may be maintained by incorporating a second phase (e.g., SiC in Al_2O_3) to inhibit grain growth (Dowding and Malghan, 1995; American Ceramic Society Bulletin, 1996). Further research is needed to determine the optimum compositions and processing parameters and to develop a commercial method for production of nanocrystalline materials.

Advanced Processing

Coatings processes developed by the electronics industry for the manufacture of semiconductor devices may help design coatings for high-temperature structural materials. For example, electronics processing relies increasingly on in situ monitoring and process diagnostics (intelligent processing) to achieve nanoscale structural control and characterization. Such techniques might improve the quality of turbine coatings and reduce the cost of applying them. In the long term, *the committee believes that intelligent processing (i.e., in situ process control technology) will be required to achieve the reproducible process control necessary for manufacture of reliable coatings.*

---

[3] Whiskers and fibers are generally very difficult to incorporate during powder processing, whereas aggregates of powders are not.

Intelligent processing of materials requires the availability and integration of process models that describe the relationships of critical process parameters with sensors that measure these parameters and appropriate control systems. Development of all these elements is required.

Built-In Sensors for Condition Monitoring

Microsensors could be embedded in TBC coatings to monitor local temperature rises, oxidation changes, and possibly incipient debonding. These sensors would monitor in real time the degradation of protective coatings and may serve to warn of imminent catastrophic failure. Alternatively, remote sensors rather than embedded sensors may achieve the same goal (NRC, 1995b).

Embedded Microchannels within TBC for Cooling

Microdesigned coatings could include small cooling channels within the TBC. TBCs could provide even better thermal insulation to the substrate by using external cooling, such as air forced along these channels. The effect of heat transfer from ceramic to coolant must be considered along with issues associated with clogging of the channels while in service. The use of lithographic patterning for formation of these channels should be considered.

To further enhance heat transport along these cooling channels, the use of steam, water, or even helium gas could be considered. To limit the amount of helium that must be carried in air-based systems, closed-loop cooling would be needed; this would be less of a concern for land-based systems. Alternatively, a fluid (gas) with poor thermal conductivity injected into the channels might serve as an insulating sheath to limit heat conduction to the TBC.

Coatings for Refractory Metals

Refractory metals are attractive, potential substrates because of their high-temperature strength and high melting temperatures. Their susceptibility to catastrophic oxidation, however, is the main barrier to their use in advanced turbine applications. The coating systems for these substrate materials must be extremely reliable since the substrate rapidly deteriorates once the coating is breached. Materials that have been explored as refractory metal coatings include polycrystalline and amorphous silicides, as well as refractory oxides such as alumina.

Further research into refractory metal coatings may yield positive results. For example, a disilicide diffusion coating for niobium could be modified with a small amount of germanium. The silica film that protects niobium from oxidation attack would be modified by the dissolution of GeO_2, which increases the expansion coefficient and lowers the viscosity of the silica glass. Thus the protective glass film could form and flow into the cracks in the coating, possibly sustaining severe temperature changes without spalling or pesting (Mueller et al., 1991).

Electron-rich noble metals (e.g., rhodium, palladium, iridium, and platinum) do not form thermodynamically stable condensed oxides and have relatively low melting temperatures. When alloyed with electron-poor metals such as hafnium, zirconium, and aluminum, however, they form remarkably stable compounds with close-packed, or nearly close-packed, structures similar to those of the structural metals themselves (Brewer, 1990).[4] These compounds have high thermodynamic stability, high melting points, and low chemical activity. For example, Brewer and Wengert (1973) report that $ZrIr_3$ melts at 2127 ± 130°C and does not interact with boiling aqua regia, molten KOH, or air at 1000°C.[5] It has the structure of $AuCu_3$, with a0 equal to 3.943 A. The formation of such alloys at the surface of a structural metal may provide a coherent, tenacious coating.

---

[4] An example is $HfPt_3$ for which the enthalpy of formation has been evaluated as $\Delta H_f^° = -132 ± 9$ kcal mole^{-1} (Brewer and Wengert, 1973).

[5] ZrI_3 has a free energy of formation of approximately ≤ -44 kcal mole^{-1} at 1527°C (Gibson and Wengert, 1984).

References

American Ceramic Society Bulletin. 1996. Technology Briefs—Nanoparticles strengthen monolithic ceramics. American Ceramic Society Bulletin 75(1):21.

Backman, D.A., and J.C. Williams. 1992. Advanced materials for aircraft engine applications. Science 255:10821087.

Bernstein, H.L. 1990. Life management system for General Electric Frame 7E gas turbine. Pp. 111-118 in Life Assessment and Repair Technology for Combustion Turbine Hot Section Components, R. Viswanathan and J.M. Allen, eds. Materials Park, Ohio: ASM International.

Biehler, D.I. 1987. Thick thermal barrier coatings for diesel engine components. Paper presented at the 1987 Proceedings of the Workshop on Coatings for Advanced Heat Engines, Castine, Maine, July 27-30. Washington D.C.: U.S. Department of Energy.

Bose, S., and J. DeMasi-Marcin. 1995. Thermal barrier coating experience in gas turbine engines at Pratt & Whitney. Pp. 63-77 in Thermal Barrier Coating Workshop. NASACP-3312. Cleveland, Ohio. National Aeronautics and Space Administration Lewis Research Center.

Boyle, R.J., and K.C. Civinskas. 1991. Two-Dimensional Navier-Stokes Heat Transfer for Rough Turbine Blades. NASA-TM-106008. Washington, D.C. : National Aeronautics and Space Administration.

Boynton, J.L., R. Tabibzadeh, and S.T. Hudson. 1992. Investigation of Rotor Blade Roughness Effects on Turbine Performance. ASME Paper 92-GT-297. New York: American Society of Mechanical Engineers.

Brewer, L. 1990. Nature of bonding in transition-metal aluminides. Journal of Physical Chemistry 94:1196-1203.

Brewer, L., and P.R. Wengert. 1973. Transition metal alloys of extraordinary stability: an example of generalized Lewis-acid-base interactions in metallic systems. Metallurgical Transactions 4:83-104.

Brindley, W.J. 1995. Properties of plasma sprayed bond coats. Pp. 189-202 in the Proceedings of the Workshop on Thermal Barrier Coating. NASA-CP-3312. Cleveland, Ohio: National Aeronautics and Space Administration Lewis Research Center.

Brindley, W.J., and J.A. Nesbitt. 1988. Advanced Earth to Orbit Propulsion Technology, Vol. 1. NASA-CP-3012. Washington, D.C.: National Aeronautics and Space Administration.

Brindley, W.J., and J.D. Whittenberger. 1993. Stress relaxation of low pressure plasma-sprayed NiCrAlY alloys. Materials Science and Engineering A 163(1):33-41.

Brindley, W.J., J.L. Smialek, and M.A. Gedwill. 1992. Oxidation resistant coatings for Ti_3Al+Nb and SiC/Ti_3Al+Nb. Pp. 41-1 to 41-15 in SITEMAP Review 1992, Vol. 2. NASA-CP-10104. Washington, D.C.: National Aeronautics and Space Administration.

Bunch, REF., ed. 1982. Deposition Technologies for Films and Coatings. Park Ridge, New Jersey: Noes.

Caccavale, C., and W. Sikkenga. 1994. Investment Casting using Core with Integral Wall Thickness Control Means. U.S. Patent, Number 5,296,308.

Cockeram, BV, G. Wang, and RA Rapp. 1995. Growth kinetics and pesting resistance of $MoSi_2$ and Ge-doped $MoSi_2$ diffusion coatings. Werkstoffe und Korrosion 46(4):207-217.

Cotton, F.A., and G. Wilkinson. 1988. Advanced Inorganic Chemistry, 5th ed. New York: John Wiley & Sons.

CRC. 1985. CRC Handbook of Chemistry and Physics, 66th ed, R.C. Weast, M.J. Astle, and W.H. Beyer, eds. Boca Raton, Florida: CRC Press.

Cruse, T.A., B.P. Johnsen, R.A. Miller, and W.J. Brindley. 1992. A presentation on thermal barrier coatings. Pp. 41-47 in the Proceedings of the 1992 Workshop on Coatings for Advanced Heat Engines, Vol. 2. Monterey, California, August 3-6. Washington, D.C.: U.S. Department of Energy.

DeMasi, J.T., K.D. Sheffler, and M. Ortiz. 1989. Thermal Barrier Coating Life Prediction Model Development Phase I, Final Report. NASA-CR-182230. Washington, D.C.: National Aeronautics and Space Administration.

DeMasi-Marcin, J.T., K.D. Sheffler, and S. Bose. 1989. Mechanisms of Degradation and Failure in a Plasma Deposited Thermal Barrier Coating. ASME Paper 89-GT-132. New York: American Society of Mechanical Engineers.

Dias, A.G., J.H. van Breda, P. Moretto, and J. Ordelman. 1995. Development of $TiN-Si_3N_4$ nano composite coatings for wear resistance applications. Journal De Physique IV 5:831.

Dimiduk, D.M., D.B. Miracle, and C.H. Ward. 1992. Overview development of intermetallic materials for aerospace systems. Materials Science and Technology 8(4):367.

REFERENCES

DOE (U.S. Department of Energy). 1992a. Clean Coal Technology Research, Development, and Demonstration: Program Plan. DOE/FE-0284. Washington, D.C.: DOE.

DOE. 1992b. Comprehensive Program Plan for Advanced Turbine Systems. DOE/FE-0279. Washington, D.C.: DOE.

Dowding, R.J., and S. Malghan. 1995. Key issues discussed at the Nanophase Materials Processing Workshop. Materials Technology 10 (9/10):95.

Doychak, J. 1994. Oxidation behavior of high-temperature intermetallics. Pp. 977-1016 in Intermetallic Compounds, Chapter 43, Vol. 1, J.H. Westbrook and R.L. Fleischer, eds. New York: John Wiley & Sons.

Duvall, D.S., and D.L. Ruckle. 1982. Ceramic thermal barrier coatings for turbine engine components. In paper presented at the 27th International Gas Turbine Conference and Exhibit, London, April 18-22. ASME 82-GT-322. New York: American Society of Mechanical Engineers.

Dyson, O.B. 1989. Modeling creep properties of coated superalloys in aggressive environments. Materials Science and Engineering 645 (11-12):A120-121.

Eaton, H.E., and R.C. Novak. 1987. Sintering studies of plasma-sprayed zirconia. Surface and Coatings Technology 32:227-236.

EPRI (Electric Power Research Institute). 1991. Guidebook and Software for Specifying High-Temperature Coatings for Combustion Turbines. EPRI GS-7334-L. Palo Alto, California: EPRI.

Evans, A.G., G.B. Crumley, and R.E. Demaray. 1983. On the mechanical-behavior of brittle coatings and layers. Oxidation of Metals 20 (5/6): 193-216.

Ferguson, B.L., G.J. Petrus, and M. Ordillas. 1994. A Software Tool to Design Thermal Barrier Coatings. Final Report. NASA Contract NAS3-2728. Washington, D.C.: National Aeronautics and Space Administration.

Fox, D.S. 1992. Environmental durability of ceramics and ceramic composites. Pp. 165-175 in Flight-Vehicle Materials, Structures and Dynamics-Assessment and Future Directions, Vol. 3, S.R. Levine, ed. New York: American Society of Mechanical Engineers.

Gabb, T.P., and R.L. Dreshfield. 1986. Superalloys data. Appendix B in Superalloys II, C.T. Sims, N.S. Stoloff, and W.C. Hagel, eds. New York: John Wiley & Sons.

Gell, M., D.N. Duhl, D.K. Gupta, and K.D. Sheffler. 1987. Advanced superalloy airfoils. Journal of Metals 39(7): 11-15.

Gibson, J.K., and P.R. Wengert. 1984. Gibbs free energies of formation for intermetallic compounds involving transition elements, lanthanides, and actinides. High Temperature Science 17:371-378.

Glassman, A.J. 1975. Turbine Design and Application, Vol. 3. NASA-SP-290-VOL-3. Cleveland, Ohio.: National Aeronautics and Space Administration Lewis Research Center.

Goward, G.W. 1987. Seventeen years of thermal barrier coatings. Paper presented at the 1987 Proceedings of the Workshop on Coatings for Advanced Heat Engines, Castine, Maine, July 27-30. Washington D.C.: U.S. Department of Energy.

Goward, G.W., and L.W. Cannon. 1988. Pack Cementation Coatings for SuperalloysHistory, Theory and Practice. ASME Paper 87-GT-50. New York: American Society of Mechanical Engineers.

Gupta, D.K, and D.S. Duvall. 1984. A Silicon and Hafnium Modified Plasma Sprayed MCrAlY Coating for Single Crystal Superalloys. Warrendale, Pennsylvania: The Materials Society.

Haafkens, M.H. 1982. Experiences in repair of hot section gas turbine components. SAE 821490. In paper presented at the Society of Automotive Engineers, Aerospace Congress and Exposition, Anaheim, California, October 25-28. Warrendale, Pennsylvania: Society of Automotive Engineers.

Hillery, R.V., B.H. Pilsner, R.L. McKnight, T.S. Cook, and M.S. Hartle. 1988. Thermal Barrier Coating Life Prediction Model Development. Final Report. NASA-CR180807. Washington, D.C.: National Aeronautics and Space Administration.

Hsu, L.L. 1987. Total corrosion control for industrial gas turbines: high temperature coatings and air, fuel and water management. Pp. 1-17 in the Proceedings of the 14th International Conference on Metallurgical Coatings, San Diego, California, March 23-27. Washington, D.C.: National Aeronautics and Space Administration.

Hsu, L.L. 1988. Total corrosion control for industrial gas turbines: air borne contaminants their impact on air/fuel/water management. Proceedings of the Gas Turbine and Aeroengine Congress, Amsterdam, The Netherlands, June 6-9. ASME 88-GT-65. New York: American Society of Mechanical Engineers.

Hsu, L.L., W.G. Stevens, and A.R. Stetson. 1979. Development and Evaluation of Processes for Deposition of Ni/Co-Cr-Al-Al-Y (MCrAlY) Coatings for Gas Turbine Components. Final Report. Contract F33615-76-C-5379. Wright-Patterson Air Force Base, Ohio: Air Force Materials Laboratory.

Ingham, H.S. and A.P. Shepard. 1965. Metco Flame Spray Handbook, Vol. III. Westbury, New York: Metco. Inc.

International Nickel Company. 1977. High Strength, High Temperature Nickel Base Superalloys, 3rd ed. Saddlebrook, New Jersey: International Nickel Company, Inc.

Jacobson, N.S. 1992. High-Temperature Durability Considerations for HSCT Combustor. NASA Technical Paper 3162. Washington, D.C.: National Aeronautics and Space Administration.

Jacobson, N.S., J.L. Smialek, and D.S. Fox. 1990. Molten salt corrosion of SiC and Si_3N_4. Pp. 99-135 in Handbook of Ceramics and Composites, Vol. 1: Synthesis and Properties, N.P. Cheremisinoff, ed. New York: Dekker.

REFERENCES

Kingery, W.D., H.K. Bowen, and D.R. Uhlmann. 1976. Introduction to Ceramics. New York: Wiley & Sons.

Lammermann, H., and G. Kienal. 1991. PVD coatings for aircraft turbine blades. Advanced Materials and Processes 140(6): 18-23.

Larsen, D.C., J.W. Adams, L.R. Johnsen, A.P.S. Teotia, and L.G. Hill. 1985. Ceramic Materials for Advanced Heat Engines. Park Ridge, New Jersey: Noyes Publications.

Lee, E.Y., and R.D. Sisson. 1994. The effect of bond coat oxidation on the failure of thermal barrier coatings: thermal spray industrial applications. Pp. 55-59 in Proceedings of the 7th National Thermal Spray Conference, Boston, Mass., June 20-24, C.C. Berndt and S. Sampath, eds. Materials Park, Ohio.: ASM International.

Lee, K.N., N.S. Jacobson, and R.A. Miller. 1994. Refractory oxide coatings on SiC ceramics. MRS Bulletin 10:35-38.

Lefebvre, A.H. 1983. Gas Turbine Combustion. Washington, D.C.: Hemisphere Publishing Corporation.

Manning-Meier, S.M., and D. Gupta. 1992. The evolution of thermal barrier coatings in gas turbine engine applications. Paper presented at the International Gas Turbine and Aeroengine Congress and Exposition, Cologne, Germany, June. ASME Paper 92-GT-203. New York: American Society of Mechanical Engineers.

Manning-Meier, S.M., D.M. Nissley, K.D. Sheffler, and T.A. Cruse. 1991. Thermal Barrier Coating Life Prediction Model Development. ASME Paper 91-GT-40. New York: American Society of Mechanical Engineers.

Maracocchi, T.F., and D.V. Rigney. 1992. PVD thermal barrier coaters: advances in physical vapor deposition coaters. GE-AE The Leading Edge (Spring).

McKee, D.W. 1993. Oxidation and protection of Ti_3Al-based intermetallic alloys. Pp. 953-958 in 1992 Materials Research Society Symposium Proceedings, Vol. 288, I. Baker, R. Darolia, J.D. Whittenberger, and M.H. Yoo, eds. Pittsburgh, Pennsylvania: Materials Research Society.

McKee, D.W., and S.C. Huang. 1990. Oxidation behavior of gamma-titanium aluminides. Proceedings of the Materials Research Society 213:939-943.

Meetham, G.W. 1988. A rational return to the stone age. Pp. 38.1-38.6 in Proceedings of the International Conference on PM Aerospace Materials, Lucerne, Switzerland, November 2-4, 1987. Shrewsbury, England: MPR Publishing Services, Ltd.

Meier, G.H., D. Appalonia, R.A. Perkins, and K.T. Chiang. 1989. Oxidation of Ti-base alloys. Pp. 185-193 in Oxidation of Intermetallics, T. Grobstein and J. Doychak, eds. Warrendale, Pennsylvania: The Materials Society.

Miller, R.A. 1984. Oxidation based model for thermal barrier coating life. Journal of the American Ceramic Society 67(8):517.

Miller, R.A. 1986. Ceramic thermal barrier coatings for electric utility gas turbine engines. Paper presented at the 3rd Berkeley Conference on Corrosion, Erosion, and Wear of Materials, Berkeley, California, January 29-31. NASATM-87288. Cleveland, Ohio: National Aeronautics and Space Administration Lewis Research Center.

Miller, R.A. 1990. Assessment of Fundamental Materials Needs for Thick Thermal Barrier Coatings for Truck Diesel Engines. NASA-TM-103130. Washington, D.C.: National Aeronautics and Space Administration.

Miller, R.A., and W.J. Brindley. 1992. Plasma sprayed thermal barrier coatings on smooth surfaces. Pp. 493-498 in Thermal Spray: International Advances in Coatings Technology. Materials Park, Ohio: ASM International.

Miller, R.A., and C.E. Lowell. 1982. Failure mechanisms of thermal barrier coatings exposed to elevated temperatures. Thin Solid Films 95 (3):265-273.

Mueller, A., G. Wang, E.L. Courtright, and R.A. Rapp. 1991. Development and cyclic oxidation behaviour of a protective $(Mo,W)(Si,Ge)_2$ coating on Nb-base alloys. Journal of the Electrochemical Society 139:1266-1277

Murphy, J.C., J.W. Maclachlan, and L.C. Aamodt. 1988. Thermal imaging of barrier coatings on refractory substrates. Review of Progress in Quantitative Nondestructive Evaluation 7:245-252

Murphy, J.C., J.W.M. Spicer, and R. Osiander. 1993. Thermal Imaging of High-Temperature Coating. Presentation to the Committee on Coatings for High-Temperature Structural Materials, National Materials Advisory Board, National Research Council, Washington, D.C., October 12.

Nealy, D., and S. Reider. 1979. Evaluation of Laminated Porous Wall Materials for Combustor Liner Cooling. ASME Paper 79-GT-100. New York: American Society of Mechanical Engineers.

Nesbitt, J.A., and R.W. Heckel. 1984. Modeling of degradation and failure of Ni-Cr-Al overlay coatings. Thin Solid Films 119:281-290.

Novak, R.C. 1994. Coating Development and Use: Case Studies. Presentation to the Committee on Coatings for High-Temperature Structural Materials, National Materials Advisory Board, National Research Council, Irvine, California, April 18-19.

NRC (National Research Council). 1986. Materials For Large Land-Based Gas Turbines. National Materials Advisory Board, Commission on Engineering and Technical Systems. NMAB-430. Washington, D.C.: National Academy Press.

NRC. 1989. On-Line Control of Metal Processing. National Materials Advisory Board, Commission on Engineering and Technical Systems. Washington D.C.: National Academy Press.

NRC. 1995a. Coal: Energy for the Future. Board on Energy and Environmental Systems, Commission on Engineering and Technical Systems. Washington D.C.: National Academy Press.

NRC. 1995b. Expanding the Vision of Sensor Materials. National Materials Advisory Board, Commission on

REFERENCES

Engineering and Technical Systems. Washington D.C.: National Academy Press.

NRC. 1995c. Computer-Aided Materials Selection for Structural Design. National Materials Advisory Board, Commission on Engineering and Technical Systems. Washington D.C.: National Academy Press.

Palko, J.E., K.L. Luthra, and D.W. McKee. 1978. Evaluation of Performance of Thermal Barrier Coatings under Simulated Industrial/Utility Gas Turbine Conditions. Final Report. DOE Contract No. EC-77-c-o5-5402. Washington, D.C.: DOE.

Pettit, F.S. 1967. Oxidation mechanisms for nickel-aluminum alloys at temperatures between 900°C and 1300°C. Metallurgical Transactions 239(September): 1296.

Ramesh, K., and A. Markworth. 1993. Functionally Graded Materials: The State of the Art. Presentation to the Committee on Coatings for High-Temperature Structural Materials, National Materials Advisory Board, National Research Council, Washington, D.C.

Rapp, R.A., and Y.S. Zhang. 1994. Hot Corrosion of Materials: Fundamental Studies. JOM 46(December):47-55.

Rigney, D.V., R. Viguie, D.J. Wortman, and W.W. Skelly. 1995. PVD thermal barrier coating applications and process development for aircraft engines. Pp. 135-150 in Proceedings of the Workshop on Thermal Barrier Coatings. NASA-CP-3312. Cleveland, Ohio: National Aeronautics and Space Administration Lewis Research Center.

Sims, C.T. 1986. Superalloys: genesis and character. Pp. 3-26 in Superalloys II, C.T. Sims, N.S. Stoloff, and W.C. Hagel, eds. New York: John Wiley & Sons.

Sims, C.T. 1991. ATS program. Advanced Materials and Processes 139(6):32-39.

Sims, C.T., N.S. Stoloff, and W.C. Hagel, eds. 1986. Superalloys II. New York: John Wiley & Sons.

Smialek, J.L. 1991. Effect of sulfur removal on Al_2O_3 scale adhesion. Metallurgical Transactions A 22(3):739-752.

Smialek, J.L. 1993. Oxidation behaviour of $TiAl_3$ coatings and alloys. Corrosion Science 35(5-8)1199-1208.

Smialek, J.L., and B.K. Tubbs. 1995. Effect of sulfur removal on scale adhesion to PWA 1480. Metallurgical and Materials Transactions A 26 (2):427.

Smialek, J.L., M.A. Gedwill, and P.K. Brindley. 1990. Cyclic oxidation of aluminide coatings on Ti_3Al + Nb. Scripta Metallurgica et Materialia 24(7):1291.

Smith, J.S., and D.H. Boone. 1990. Platinum-Modified Aluminides-Present Status. Paper presented at the International Gas Turbine and Aeroengine Congress and Exposition, Brussels, Belgium, June.

Soechting, F. 1994. Gas Turbine Design Issues. Presentation to the Committee on Coatings for High-Temperature Structural Materials, National Materials Advisory Board, National Research Council, Washington D.C., February 17-18.

Soechting, F. 1995. A design perspective on thermal barrier coatings. Paper presented at the Proceedings of the Workshop on Thermal Barrier Coating. NASA-CP-3312. Cleveland, Ohio: National Aeronautics and Space Administration Lewis Research Center.

Sprague, R., and S. Freisen. 1986. Superalloy component durability enhancements. Journal of Metals 38(7):24-30.

Stephens, J.R. 1993. Materials/Coatings Requirements for the HSCT Propulsion System. Presentation to the Committee on Coatings for High-Temperature Structural Materials, National Materials Advisory Board, National Research Council, San Antonio, Texas, December 13-14.

Stoloff, N.S., and C.T. Sims. 1986. Alternative materials. Pp. 519-547 in Superalloys II, C.T. Sims, N.S. Stoloff, and W.C. Hagel, eds. New York: John Wiley & Sons.

Strangman, T.E. 1985. Thermal barrier coatings for turbine airfoils. Thin Solid Films 127:93-105.

Strangman, T.E. 1987. Development and performance of physical vapor deposition thermal barrier coating systems. Paper presented at the 1987 Proceedings of the Workshop on Coatings for Advanced Heat Engines, Castine, Maine, July 27-30. Washington, D.C.: U.S. Department of Energy.

Strangman, T.E. 1990. Turbine coating life prediction model. In the 1990 Proceedings of the Workshop on Coatings for Advanced Engines, August. Washington, D.C.: U.S. Department of Energy.

Stringer, J., and R. Viswanathan. 1990. Keynote address: Life assessment techniques and coating evaluations for combustion turbine blades. Pp. 1-18 in Life Assessment and Repair Technology for Combustion Turbine Hot Section Components: Proceedings of an International Conference. Materials Park, Ohio: ASM International.

Stringer, J., and R. Viswanathan. 1993. Gas turbine hot-section materials and coatings in electric utility applications. Pp. 1-21 in Proceedings of ASM 1993 Materials Congress Materials Week '93, Pittsburgh, Pennsylvania, October 17-21. Materials Park, Ohio: ASM International.

Taguchi, G. 1993. Robust Development. New York: American Society of Mechanical Engineers Press.

Tiegs, T.N., M.K. Ferber, and P.F. Becher. 1992. Toughened ceramic composites: second-phase particulates, transformation toughening and whisker reinforcement. Pp. 43-57 in Flight-Vehicle Materials, Structures and Dynamics Assessment and Future Directions, Vol. 3, S.R. Levine, ed. New York: American Society of Mechanical Engineers.

Turner, E., J. Rhodes, and L. Junod. 1992. Cooled Wall Structure Especially for Gas Turbine Engines. U.S. Patent, Number 5,152,667.

Ulion, N., and N. Anderson. 1993. Advanced Thermal Barrier Coated Superalloy Components. U.S. Patent, Number 5,262,245.

Van Roode, M., J.R. Price, and C. Stala. 1992. Ceramic oxide coatings for the corrosion protection of silicon carbide. Journal of Engineering Gas Turbines Power Transactions ASME 115(1):139-149.

REFERENCES

Vasudevan, A.K., and J.J. Petrovic. 1992. A comprehensive overview of molybdenum disilicide composites. Materials Science and Engineering A155(1-2): 1-17.

Walston, W.S., E.W. Ross, K.S. O'Hara, and T.M. Pollock. 1993. U.S. Patent, Number 5,270,123.

Wood, J.H., and E.H. Goldman. 1986. Protective coatings. Pp. 359-384 in Superalloys II, C.T. Sims, N.S. Stoloff, and W.C. Hagel, eds. New York: John Wiley & Sons.

Yonushonis, T.M., D.P. Roehling, K.L. Hoag, R.C. Novak, A.P. Matarese and R.P. Huston. 1987. Thick thermal barrier coatings for diesel engines. Paper presented at the 1987 Proceedings of the Workshop on Coatings for Advanced Heat Engines, Castine, Maine, July 27-30. Washington D.C.: U.S. Department of Energy.

REFERENCES

Appendix A

Testing and Standards

Broadly accepted test methods, standards, and specifications are of great value to both vendors and purchasers of coating services and coated products. They are essential communication mechanisms for purchasers to describe the critical aspects of required coatings and for coating suppliers to unambiguously understand the requirements.

Coatings are only used in gas turbines after a substantial testing and evaluation program by the engine manufacturer and, where appropriate, by the coating vendor. Testing and analysis procedures are usually developed by the engine manufacturer to:

- develop procedures, specifications, and controls for the deposition process
- serve as a quality control measure to ensure that coated products meet specified properties
- provide data for performance and lifetime prediction by evaluation under conditions simulating engine conditions

These procedures are developed by engine manufacturers to address specific conditions expected in each manufacturer's engine. Notable examples of this approach are found in the several types of high-temperature tests developed by engine manufacturers to simulate the corrosive behavior found in specific engines operating under specific conditions. Burner rigs of various designs have evolved at each manufacturer that, through experience, can generate data for corrosion and thermal-shock resistance. Many different methods of testing and analysis have evolved in the research community to develop new coating compositions, microstructures, and processes as well as to understand coating behavior on a fundamental level.

The measurement of coating properties must be viewed in the context of a coating/substrate system. Coating compositions and microstructures are complex and become more so during service at high temperature in corrosive environments. Thus the use of material properties for design or lifetime prediction is usually based on the measurement of the coating systems rather than the bulk materials, the properties of which may differ significantly from the same nominal materials present as a coating.

While substantial efforts have been made by individual companies and research organizations to develop satisfactory evaluation procedures, relatively few broadly accepted test methods are available for the evaluation of high-temperature coatings. Consequently, much of the data publicly reported consist of measurements that are directly compared with the behavior of widely used materials. This is a conservative approach suitable for the expensive turbines that are expected to have high reliability. However, the need for increased productivity in the materials and gas-turbine fields argues for the use of commonly accepted test methodologies that allow more cost-effective data generation and increased commonality of property specification.

Definitions for the several terms used for standards have been developed by the American Society for Testing and Materials (ASTM) and are followed in this discussion. A standard is defined as a rule for an orderly approach to a specific activity, formulated and applied for the benefit and with the cooperation of all concerned. Six types of full consensus standards are identified by the ASTM:

- Classification. A systematic arrangement or division of materials, products, systems, or services into groups based on similar characteristics (e.g., origin, composition, properties, or use).
- Guide. A series of options or instructions that does not recommend a specific course of action.
- Practice. A definitive procedure for performing one or more specific operations or functions that does not produce a test result.
- Specification. A precise statement of a set of requirements to be satisfied by a material, product, system, or service that also indicates the procedures for determining whether each of the requirements is satisfied.
- Terminology. A definition or description of terms or an explanation of symbols, abbreviations, or acronyms.
- Test method. A definitive procedure for the identification, measurement, and evaluation of one or more qualities, characteristics, or properties of a material, product, system, or service that produces a test result.

In general, company specifications dominate the commercial market and address characteristics such as composition, microstructure, thickness, and strain-to-failure. The U.S. military has published specifications that address limited aspects of high-temperature coatings, primarily for the thermal spray (plasma and detonation gun) processes. Broadly available standards developed by consensus through organizations such as the Society of Automotive Engineers and ASTM address feedstock composition and powder size for thermal spray processes with limited coating property measurements. Foreign standards, notably British and German, primarily address coating thickness with limited attention to physical or mechanical properties.

Testing standards are particularly important for thermal barrier coatings (TBCs) in that they include a ceramic layer(s) and the inherent scatter in the mechanical properties of ceramics is accentuated by the complex microstructure produced by thermal spraying or electron-beam physical vapor deposition (EB-PVD). The lack of standard test methods and data analysis and interpretation techniques for relatively fundamental properties (e.g., strength, adhesion and cohesion, strain-to-failure, and ductility) is accompanied, not unexpectedly, by a lack of standards for more complex properties (e.g., thermal shock, fatigue, wear and erosion, corrosion, and toughness). Although basic and applied research has been conducted to understand coating behavior and to relate processing and microstructure and microchemistry to properties and performance, little of this effort has resulted in standards. The necessity to determine accurately appropriate properties for thermal-spray-deposited coatings has been recognized (Berndt et al., 1992; Dapkunas, 1993), but a similar perspective for coatings applied by other processes has not been developed. For TBCs, the lack of understanding of failure mechanisms hinders the identification of required standards. This situation that exists for current superalloy components is also present for future materials such as monolithic ceramics and ceramic-matrix composites, which may require coatings for oxidation protection.

The measurement of properties for use in coating micromechanical design is a significant area that has been neglected. Measurements of properties (e.g., modulus of elasticity, coefficient of thermal expansion (CTE), inter-and intragranular strength, and toughness) would be particularly valuable for design of TBCs and the functionally graded materials that are similar in concept. The difficulty of measurement on the micrometer scale required for these materials is largely responsible for this situation. Test techniques currently in development (e.g., nanoindentation) may alleviate this situation.

High-temperature coatings have not been the specific subject of standards development, and many of the standards developed for other applications have been used where appropriate. The status of standard testing and analysis procedures varies with the specific aspect of coating technology considered and is summarized in the following section.

NOMENCLATURE AND DESIGNATION

The description of high-temperature coating types and coating processes lacks well-defined, universally accepted terminology. For example, the same process may be described as *pack aluminizing* or *chemical vapor deposition.* Similar ambiguity occurs in the use of the term *thermal spraying,* which may inclusively refer to all high-temperature, gas-propelled particulate applications to a substrate or specifically to high-velocity, oxygen-fueled deposition. The proprietary, rather than commodity, nature of the coatings industry necessitates that a specific coating be designated by the manufacturer rather than by technical or trade bodies. There are clear benefits for using commonly accepted terms to describe coating processes or attributes, but there does not appear to be value in the development of commonly accepted coating designations.

PROCESSING

Standards, or more precisely procedures, for processing are generally developed by coating producers and users. However, the reliance of the military on coatings to achieve desired performance objectives has fostered the issue of specifications for both plasma spray and detonation gun deposition processes. ASTM standards developed for the minerals-handling and powder metallurgy industries have been adopted for use where applicable by the plasma spray industry for the analysis of powders. Although company specifications for processes (e.g., EB-PVD or sputter deposition) are used, they have not been adopted as publicly available standards. Furthermore, suitable reference materials are not available for the calibration of analytical instruments necessary for process control (Dapkunas, 1993).

MECHANICAL PROPERTIES

Mechanical properties, particularly as a function of temperature, are critical to the performance of a coating and consequently are the subject of measurement in the coatings field.

The properties of greatest concern for metallic overlay or *conversion* coatings are ductile-to-brittle transition, fatigue, thermal-shock resistance, adhesion, and strain-to-failure. Test procedures for these properties are generally of the elevated-temperature uniaxial tensile, creep, stress-rupture, or fatigue type developed for metals and alloys. The evaluation of ceramic overlay coatings used for thermal barriers focuses on adhesion to the substrate and cohesion within the coating.

Traditional methods for the qualitative evaluation of adhesion (e.g., bending, scratching, or impacting) that were developed for less complex materials (e.g., zinc coatings) are of limited value but are nonetheless included in ASTM, British Standards Institute, and International Standards Organization (ISO) standards. The most commonly accepted adhesion test is ASTM C 633-79, the tensile tab test, which is comparable to DIN 50 160-A, AFNOR NF A91-202-79, and JIS H866680 (Berndt, 1990). This technique is limited by the use of an epoxy adhesive grip attachment at test temperatures significantly lower than service temperatures. Brown et al. (1988) have provided a review of methods used to measure the adherence of coatings applied by thermal spray, including flexure and fracture mechanics techniques, and conclude that widely used tests do not provide the information required and that simulation of service conditions is vital. Thermal-shock tests can provide a qualitative measure of adhesion and have been developed for the porcelain-coated steel industry. The importance of determining the mechanical properties of coatings has encouraged the development of compressive, tensile fatigue techniques for coatings removed from substrates (Beardsley, 1992). These methods have not been codified as standards.

Recognition of the importance of more subtle properties such as fracture toughness, thermal-shock response, and thermomechanical fatigue has been manifested in research on the modeling of coating behavior and was the subject of a recent conference (Kokini, 1993). However, standards for the measurement of these properties are not available.

Hardness is commonly used as a process control measure, and its application to coatings has been recognized in the development of BS 5411-part 6, Vickers and Knoop microhardness, for metallic coatings. Fracture mechanics analysis has been combined with microindentation to measure the fracture toughness of ceramics and has been applied to ceramic coatings (Besich et al., 1993). More recently, instrumented microindenters and nanoindenters that provide data on deformation as a function of load have been developed that can provide a measure of elastic properties. Nanoindentation offers the potential to measure hardness and elastic modulus of specific portions of microstructures as small as several micrometers that can be applied to modeling. None of these latter techniques have been developed into standards.

CORROSION

Metallic overlay and conversion coatings are used to impart oxidation and hot-corrosion resistance to alloy substrates. Ceramic TBCs are expected to exhibit corrosion resistance and to protect the substrate as well.

Corrosion behavior is determined by a variety of static and dynamic tests in environments selected to simulate expected, pertinent aspects of turbine-operating environments. Static tests in furnaces with stagnant or low gas flows provide information on thermodynamic stability and reaction kinetics. Data usually consist of weight changes with time, corrosive penetration determination by metallographic examination, phase-change identification, and corrosive product formation.

Dynamic testing is usually conducted in burner rigs operating at atmospheric pressure and gas velocities of less than 100 feet per second. High-velocity and elevated-pressure burner rigs that more closely simulate turbine conditions are used less frequently. Specimen analysis of burner-rig samples usually consists of metallographic examination and compositional analysis of corrosion products. Standards for the conduct of these tests, the specimen types, and the interpretation of data are not available. Nonetheless, significant amounts of research have allowed the comparison of data among the different tests.

Recognition of the value of understanding the relationships among the test methodologies for hot corrosion is reflected in the early publication of ASTM STP-421 that provides focused coverage of the tests used in the 1960s (ASTM, 1967). Similar comparisons of more recently developed tests have not been published. The evaluation of corrosion and thermal-shock resistance of TBCs, particularly plasma-sprayed coatings, has been the subject of considerable research. Engine manufacturers and the NASA Lewis Research Center have used burner rigs of various designs for this purpose, and research at the latter institution has highlighted the necessity of well-designed experiments (Miller et al., 1993).

EROSION

Although test methodologies to evaluate the erosion of coatings and substrate alloys in turbine environments have been used extensively, standards for generally accepted techniques have not been developed. ASTM G76-83 (Standard Practice for Conducting Erosion Tests by Solid Particle Impact Using Gas Jets), for example, uses large particles at low velocities at temperatures of 18-28°C. Typically, evaluation tests for turbine materials are conducted in burner rigs that use particle injection or in high-temperature furnaces using particles entrained in high-velocity gas streams. Specimen evaluation usually consists of a measurement of surface recession and the data provide a ranking of the materials examined. The degree of erosion damage is influenced by particle properties (e.g., size and hardness), impact angle, and velocity. These influences are in addition to coating properties (e.g., hardness and microstructure) that may vary with temperature and long-term exposure to operating conditions. These parametric difficulties may render standard erosion tests that provide data for accurate performance prediction an unrealistic expectation.

THERMAL PROPERTIES

Thermal properties are a particular concern for ceramic TBCs. Knowledge of thermal conductivity, preferably at the temperature of intended use, is important for design and life prediction, while knowledge of thermal expansion coefficients are critical to understanding the adherence to the substrate and stresses in the coating.

Thermal conductivity of bulk materials and coatings is routinely measured by the use of a laser flash apparatus that provides a controlled heat input to the front of a sample and measures surface temperature change at the back. Direct measurements are obtained through the use of the guarded hot plate technique that uses a well-insulated apparatus to heat one side of a specimen, while the temperature is measured on the other side. Standards for the guarded hot plate technique (ASTM C-177) and the laser flash (ASTM E-1461) are available for uncoated specimens and may be extended to coated specimens. The ease of the laser flash technique makes it attractive for commercial use and argues for the development of a standard reference material that can be related to results from the guarded hot plate method.

Heat transfer through coatings is influenced by the thermal emissivity as well as the thermal conductivity of the coating. Techniques for measuring emissivity are available and will become more important as operating temperatures increase. These techniques are not available as standards for coatings.

The thermal comparator method has been used to determine the conductivity of films thinner than 1 micron. This work has shown that these materials can exhibit conductivities as much as two orders of magnitude lower than bulk materials of the same composition and that significant interfacial thermal resistance can develop (Lambropoulos et al., 1993). This observed behavior has implications for the use of thin multilayer TBCs and warrants the evaluation of this class of materials by techniques for which standard methods are available.

CTEs can be measured on coatings removed from a substrate using conventional dilatometry, but the graded composition and microstructure of TBCs adds a degree of complexity that can result in specimen bowing. Although bowing may complicate conventional measurements, this phenomena could, in principle, be used as an alternative method to determine coating expansion. Standards for determining the CTE for coatings are unavailable.

MICROSTRUCTURE

Microstructure of coatings is routinely inspected and analyzed by metallographic preparation of cross sections. Although examination of metallic overlay and conversion coatings is relatively straightforward, graded coatings, which can include porosity and relatively loosely bonded material in the case of plasma-sprayed material, presents more difficulty. Special techniques (e.g., epoxy infiltration) are required to preserve the coating microstructure. Procedures for metallographic preparation and microstructural analysis have been developed by coating producers and users but have not been codified as standard methods.

Porosity, manifested as coating density, has significance for coating thermomechanical behavior, corrodent penetration, heat transfer, and centrifugal loading of turbine disks. Standards for measurement of connected porosity by use of the BET gas absorption technique are routinely used (e.g., ASTM C-577-92-refractory permeability). Unconnected porosity is more difficult to measure. Sophisticated techniques (e.g., small-angle neutron scattering) have been used in research to determine pore size and distribution. This technique is not suitable for routine laboratory or production use but may provide a means to synthesize standard materials for calibration of metallographic or other techniques.

INTERNATIONAL ACTIVITIES

Although the lack of standards for the evaluation of coatings has not impeded the development of new materials for gas turbines, the increased emphasis on international trade may well cast standards in a new light. The greater interest in standards developed by national standards organizations is manifested for materials generally by activities conducted under the auspices of the Versailles Agreement on Materials and Standards that addresses prestandards research. Standards for coatings are addressed specifically by the Committee for European Normalization Technical Committee 184 for Advanced Technical Ceramics (Working Group 5-ceramic coatings). These standards generally correspond to the standards set by national standards organizations. Nations that do not participate in the development of these standards are placed in the unenviable position of having to provide test data according to procedures developed by competitors. Although this is not an insurmountable obstacle to overseas marketing, the early presence of domestic firms in the development of these standards allows the specific concerns of domestic firms to be accorded consideration and shape the standards finally adopted. Participation of U.S. firms in this activity is clearly important for the long-term well-being of the domestic turbine and coatings industries. The development of standards for analysis of coatings can have immediate, positive effects in providing data that are more useful to the turbine industry, but long-term changes that generally affect industry should also be considered.

ISO 9000, an international standard for certification that manufacturers have implemented quality assurance procedures, is increasingly required for sales of products. One facet of this procedure is the identification of test and analysis procedures that ensure that products have specified properties. The availability of standards for coating evaluation provides well-accepted criteria for use in the certification procedure. Similarly, for many coating processes, reproducible deposition relies on the skill and judgment of equipment operators. Certification of these operators, similar to that required for welders, may be desirable in meeting the requirements of ISO 9000. This requirement has fostered the formation of the National Aerospace and Defense Contractors Accreditation Program, which aims to develop industrywide quality accreditation procedures.

The increased exchange of product data among manufacturers has encouraged the development of computer protocols for the exchange of materials information. This activity is conducted under the auspices of ISO Technical Committee 104, subcommittee 4, and is identified informally as STEP (Standard for the Exchange of Product model) and formally as ISO 10303. Completion of the program is expected by the year 2000. Although originally focused on the exchange of data for computer-aided design, a program is in place to allow the exchange of material data from producers to users (Rumble and Carpenter, 1992). In 1994, standards drafted by the committee for the description of materials properties were developed and are expected to be adopted as ISO standards. Application protocols for the exchange of data relative to specific industries are planned, with testing of polymers the likely first use. International commerce in turbine coatings could well be impacted by this data exchange methodology and warrants participation by the domestic turbine and coating manufacturers.

SUMMARY

Relatively few standard test and analysis techniques have been developed for high-temperature coatings. The coatings community would be well served by the availability of broadly accepted standard measurement techniques for microstructure, mechanical, and thermal properties on scales appropriate for the design and analysis of coating systems. The Committee for European Normalization has identified coatings measurement standards as a topic for development, and it is reasonable to expect that these standards will be incorporated into the ISO code. The interests of producers and users of domestic coatings would be well served by the participation of these parties in the development of standards that reflect their methods of measurement. The current standards for coatings are listed below.

CURRENT STANDARDS

American Society for Testing and Materials

Processing	
B 212-89	Standard Test Method for Apparent Density of Free-Flowing Metal Powders
B 213-90	Standard Test Method for Flow Rate of Metal Powders
B 214-92	Standard Test Method for Sieve Analysis of Granular Metal Powders
B 215-90	Standard Test Methods of Sampling Finished Lots of Metal Powders
C 702-87	Standard Practice for Reducing Field Samples of Aggregate to Testing Size
Properties	
B 571-91	Standard Test Methods for Adhesion of Metallic Coatings (bend, burnishing, chisel-knife, draw, file, grind-saw, heat-quench, impact, peel, push, scribe-grid tests, qualitative only; similar to BS 5411/Part 10)
C 177-85	Steady State Heat Flux Measurements and Thermal Transmission Properties by Means of the Guarded-Hot-Plate Apparatus
C 313	Adherence of Porcelain to Steel (discontinued)
C 633-79	Standard Test Method for Adhesion or Cohesive Strength of Flame Sprayed Coatings (reapproved 1993)
C 577-92	Standard Test Method for Permeability of Refractories
Other	
D 4541-85	Standard Test Method for Pull-Off Strength of Coatings Using Portable Adhesion Testers
F 692-80	Standard Test Method for Measuring Adhesion Strength of Solderable Films to Substrates (reapproved 1991)

APPENDIX A

Society of Automotive Engineers

Processing

AMS 2435C	Detonation Process-Tungsten Carbide/Cobalt Coating
AMS 2436B	Coating, Aluminum Oxide-Detonation Deposition
AMS 2437B	Coating, Plasma Spray Deposition
AMS 5791	Powder, Plasma Spray, 56. 5Co-25. 5Cr-10. 5Ni-7. 5W
AMS 5792	Powder Plasma Spray, 50(88W-12Co) + 35(70Ni-16.5Cr-4Fe-4Si-3.8B) + 15(80Ni-20Al)
AMS 5793	Powder, Plasma Spray, (95Ni-5A1)
AMS 7875	Chromium Carbide Plus Nickel-Chromium Alloy Powder, 75Cr2C3 + 25 (80Ni-20Cr Alloy)
AMS 7878	Tungsten Carbide Powder, Cobalt Coated
AMS 7879	Tungsten Carbide-Cobalt Powder, Cast and Crushed
AMS 7880	Tungsten Carbide-Cobalt Powder, Sintered and Crushed

Military Standards

Processing

MIL-C-52023	Coating: Ceramic, Refractory, for High Temperature Protection of Low Carbon Steel -1958
MIL-STD-1886	(AT) Tungsten Carbide-Cobalt Coating, Detonation Process -1992
MIL-C-81751B	Coating, Metallic Ceramic
MIL-M-80141C	Metallizing Outfits, Powder Guns and Accessories -1987
MIL-STD-1884A	(AT) Coating, Plasma Spray Deposition -1991
MIL-Z-81572	(AS) Zirconium Oxide, Lime Stabilized, Powder and Rod, for Flame Spraying -1991
MIL-83348	Powders, Plasma Spray (CANCELED)

British Standards Institute

Properties

BS 5411	Methods of Test for Metallic and Related Coatings
Part 1	Definitions and Conventions Concerning the Measurements of Thickness
Part 2	Review of Methods for the Measurement of Thickness
Part 3	Eddy Current Method for Measurement of Thickness of Non-Conductive Coatings on Non-Magnetic Basis Materials
Part 4	Coulometric Method for the Measurement of Coating Thickness
Part 5	Measurement of Local Thickness
Part 6	Vickers and Knoop Microhardness Tests
Part 7	Profilometric Method for Measurements of Coating Thickness
Part 8	Measurement of Coating Thickness of Metallic Coatings: X-Ray Spectrometric Methods
Part 9	Measurement of Coating Thickness of Electrodeposited Nickel Coatings on Magnetic and Non-Magnetic Substrates-Magnetic Method
Part 10	1981/ISO 2819-1980 Review of methods for testing adhesion of electrodeposited and chemically deposited metallic coatings on metallic substrates (burnishing, ball burnishing, shot peening, peel [less than 125 microns], file, grinding and sawing, chisel [greater than 125 microns], scribe and grid, bending, twisting, tensile, thermal shock, drawing, cathodic; qualitative only)
Part 11	Measurement of Coating Thickness of Non-Magnetic Metallic and Vitreous or Porcelain Enamel Coatings on Magnetic Basis Metals: Magnetic Method
Part 12	Beta Backscatter Method for Measurement of Thickness
Part 13	Chromate Conversion Coatings on Zinc and Cadmium

APPENDIX A

Part 14	Gravimetric Method for Determination of Coating Mass per Unit Area of Conversion Coatings on Metallic Materials
Part 15	Review of Methods of Measurement of Ductility
Part 16	Scanning Electron Microscopy Method for Measurement of Local Thickness of Coatings by Examination of Cross Sections
BSI M. 40	Aerospace Series-Methods for Measuring Coating Thickness by Non-Destructive Testing

DIN

Processing

50961-87	Electroplated Coatings: Zinc and Chromate Coatings on Iron and Steel: Chromate Treatment of Zinc and Cadmium Coatings
50966-88	Electroplated Coatings: Autocatalytic Nickel-Phosphorus Coatings on Metal in Technical Applications
50967-91	Electrodeposited Coatings of Nickel Plus Chromium and Copper Plus Nickel Plus Chromium
50968-91	Electrodeposited Coatings of Nickel and Nickel Plus Copper

Properties

50933-87	Measurement of Coating Thickness by Differential Measurement Using a Stylus
50949-84	Non-Destructive Testing of Anodic Oxidation Coatings on Pure Aluminum and Aluminum Alloys by Measurement of Admittance
50955-83	Measurement of Coating Thickness. Measurement of Thickness of Metallic Coatings by Local Anodic Dissolution: Coulometric Method
50976-89	Corrosion Protection: Hot Dip Batch Galvanizing: Requirements and Testing
50978-85	Testing of Metallic Coatings: Adherence of Hot Dip Zinc Coatings [up to 150 microns]
50982 PT	1-87 Principles of Coating Thickness Measurement: Terminology Associated with Coating Thickness and Measuring Areas
50982 PT	2-87 Principles of Coating Thickness Measurement: Review of Commonly Used Methods of Measurement
50982 PT	3-87 Principles of Coating Thickness Measurement: Selection Criteria and Basic Measurement Procedures
50987-87	Measurement of Coating Thickness by the X-Ray Spectrometric Method
50160-A	Tensile Adhesion

Other

50960 PT	1-86 Electroplated and Chemical Coatings: Designation and Specification in Technical Documents
50960 PT	2-86 Electroplated and Chemical Coatings: Indications on Drawings

Japan Institute of Standards

Properties

H8666-90	Testing Methods for Thermal Sprayed Ceramic Coatings
R4204	Method of Testing Ceramic Coating

APPENDIX A

REFERENCES

ASTM (American Society for Testing and Materials). 1967. Hot Corrosion Problems Associated with Gas Turbines. Special Technical Publication 421. Philadelphia, Pennsylvania: ASTM.

Beardsley, M.B. 1992. Thick thermal barrier coatings. Pp. 567-572 in Proceedings of the Annual Automotive Technology Development Contractors' Coordination Meeting, Dearborn, Michigan, November 2-5. Warrendale, Pennsylvania: Society of Automotive Engineers.

Berndt, C.C. 1990. Tensile adhesion testing methodology for thermally sprayed coatings. Journal of Materials Engineering 12:151-158.

Berndt, C.C., W. Brindley, A.N. Goland, H. Herman, D.L. Houck, K. Jones, R.A. Miller, R. Neiser, S. Sampath, M. Smith, and P. Spanne. 1992. Current problems in plasma spray processing. Journal of Thermal Spray Technology 1(4):341-356.

Besich, G.K., C.W. Florey, F.J. Worzala, and W.J. Lenling. 1993. Fracture toughness of thermal spray ceramic coatings determined by the indentation technique. Journal of Thermal Spray Technology 2(1):35-38.

Brown, S.D., B.A. Chapman, and G.B. Wirth. 1988. Fracture kinetics and the mechanical measurement of adherence. Pp. 147-157 in Proceedings of the National Thermal Spray Conference, October 24-27, D. Hauck, ed. Metals Park, Ohio: ASM International.

Dapkunas, S.J. 1993. NIST-Industry Workshop on Thermal Spray Coatings Research. NIST Journal of Research 98(3):383-389.

Kokini, K., ed. 1993. Ceramic coatings. In Proceedings of the 1993 ASME Winter Annual Meeting, Materials Division, New Orleans, Louisiana, November 28-December 3. New York: American Society of Mechanical Engineers.

Lambropoulos, J.C., S.D. Jacobs, S.J. Burns, L. ShawKlein, and S.S. Hwang. 1993. Thermal conductivity of thin films: measurement and microstructural effects. Pp. 21-32 in Thin Film Heat Transfer-Properties and Processing, Vol. 184, M.K. Alam, M.I. Slik, G.P. Grigoropoulos, J.A.C. Humphrey, R.L. Mahajan, and V. Prasad, eds. New York: American Society of Mechanical Engineers.

Miller, R.A., G.W. Lissler, and J.M. Jobe. 1993. Characterization and Durability of Plasma Sprayed Zirconia-Yttria and Hafnia-Yttria Thermal Barrier Coatings. NASA Technical Paper 3295. Washington, D.C.: National Aeronautics and Space Administration.

Rumble, J., and J. Carpenter. 1992. Materials STEP into the future. Advanced Materials and Processes 142(4):23-27.

Appendix B

Radiation Transport in Thermal Barrier Coatings

The energy transport process in ceramic thermal barrier coatings (TBCs) has been measured until recently by an empirical parameter: thermal conductivity. This measure does not distinguish the relative contributions of radiation and true conduction. A role for radiative transport is raised by the transparency of some of the ceramics used for TBC applications in the infrared spectral region for wavelengths in the region of 1 to 6 µm (Thomas and Joseph, 1987; Thomas, 1989; Sova et al., 1992). This region also corresponds to the region of maximum black-body emission for the operating temperatures of present gas-turbine engines. Advanced engines under development in current NASA and U.S. Department of Energy programs are expected to reach even higher temperatures with correspondingly higher levels of infrared radiation incident on the coating in the 1- to 6-µm band from the interior of the engine. The combination of high infrared radiation density and low coating absorption suggests that radiation may compete with conduction for energy transport across the coating.

Radiative transport of energy across a uniform coating involves a number of factors:

- the intensity and spectral distribution of infrared radiation in the engine enclosure in the 1- to 6-µm region
- infrared reflection at the coating/engine and coating/blade interfaces
- scattering and absorption within the coating
- radiation trapping in the coating by internal reflection of radiation

Regarding the source energy distribution, the radiant energy that is incident on a coated blade or vane in the engine will have contributions from both the burning gas and the other engine surfaces. The relative contributions of these sources may change with position in the engine and blade number. Factors influencing infrared reflection at the coating/engine interface include a quasi-specular component associated with the difference in dielectric constants of the gas and coating as well as a diffuse reflection associated with the structure of the coating surface. For example, coatings applied using an electron-beam physical vapor deposition (EB-PVD) process have a polycrystalline columnar structure that leads to significant scattering in the visible spectral region. The same coatings are partially transparent in the infrared in the 1- to 6-µm region (Murphy et al., 1993), however, suggesting that scattering is size related via the parameter ka where a is a dimension of the crystallites in the coating and $K=2\pi D/\lambda D$ is the characteristic size of the scatterer, and λ is the wavelength. In the case of EB-PVD coatings the scatters are features such as the spacing and organization of columns in the coating (the coating has a columnar structure) and individual crystallites in a column. Both of these contribute to the scattering. This result further implies that the penetration of radiation from an external source will depend on the wavelength and angle of incidence.

Scattering has been shown to play an important role in determining the internal temperature distribution in layered specimens (Siegel and Spuckler, 1993b). For specimens that exhibit both scattering and absorption of radiation, the temperature distribution is affected by the albedo, which is a measure of the relative importance of scattering and absorption (Siegel and Spuckler, 1993a). Radiation trapping caused by internal reflection was also shown to play an important role in determining the temperature distribution (Siegel and Spuckler, 1993a). In light of the roles that scattering and radiative transport play in reducing the temperature gradient (i.e., creating more uniform temperatures along the blade) and since temperature gradient drives thermal conduction, any estimate of the relative importance of radiation and conduction requires that a treatment consider both mechanisms simultaneously.

TBCs fabricated from layers of oxides of varying optical thickness have been proposed to reduce radiative transport by selective reflection of infrared radiation in the 1- to 6-µm band (Soechting, 1994). These TBCs would form high-temperature interference filters in the near infrared. In addition, multilayered materials will affect the rate of conductive transport (Aamodt et al., 1990a,b). For example, multilayered structured TBCs, or more generally nanostructured TBCs, could reduce both conductive and radiative transport. The relative importance of the two processes should be determined through more extensive study.

REFERENCES

Aamodt, L.C., J.W.M. Spicer, and J.C. Murphy. 1990a. Analysis of characteristic thermal transit times for time resolved infrared radiometry studies of multilayered coatings. Journal of Applied Physics 68(12):6087-6098.

Aamodt, L.C., J.W.M. Spicer, and J.C. Murphy. 1990b. The effect of transverse heat flow and the use of characteristic times in studying multilayered-coatings in the time domain. Pp. 59-63 in Photoacoustic and Photothermal Phenomena II, J.C. Murphy, J.W.M. Spicer, L.C. Aamodt, and B.S.H. Royce, eds. Berlin, Germany: Springer-Verlag.

Murphy, J.C., J.W.M. Spicer, and R. Osiander. 1993. Thermal Imaging of High-Temperature Coating. Presentation to the Committee on Coatings for High-Temperature Structural Materials, National Materials Advisory Board, National Research Council, Washington, D.C., October 12.

Siegel, R., and C.M. Spuckler. 1993a. Variable refractive index effects on radiation in semitransparent scattering multilayered regions. Journal of Thermophysics and Heat Transfer 7:624-630.

Siegel, R., and C.M. Spuckler. 1993b. Refractive index effects on radiation in an absorbing, emitting and scattering laminated layer. Transactions of American Society of Mechanical Engineers 115:194-200.

Soechting, F. 1994. Gas Turbine Design Issues. Presentation to the Committee on Coatings for High-Temperature Structural Materials, National Materials Advisory Board, National Research Council, Washington D.C., February 17-18.

Sova, R., M.J. Linevsky, M.E. Thomas, and F.F. Mark. 1992. High temperature optical properties of oxide ceramics. Johns Hopkins APL Technical Digest 13:369-378.

Thomas, M.E. 1989. A computer code for modeling optical properties of window materials. Proceedings of SPIE: Window and Dome Technologies and Materials 1112:260-267.

Thomas, M.E., and R.I. Joseph. 1987. Characterization of the complex index of refraction for sapphire, spinel, alon and yttria. Proceedings of the IRIS Specialty Group on Infrared Materials, June 9-10.

Appendix C

Survey of Nondestructive Evaluation Methods

For coating systems in high-temperature gas-turbine applications, the primary nondestructive evaluation (NDE) issues are the ability to measure the stability and adherence of materials at elevated temperatures. Processes such as oxidation, hot corrosion, creep, and solid-state diffusion degrade performance of both coating and substrate at high temperatures with relatively high reaction rates. These processes can lead to spallation of the coating, coating/substrate interdiffusion, and crack formation in the coating or substrate. These changes can occur at the interface between coating and substrate or at other interior sites. Consequently an important criterion for selecting an NDE method is the ability to monitor the subsurface condition of coated structures. However, in view of the many goals for NDE and the wide variety of coating systems, failure modes, and component designs, no single NDE approach can meet all needs. Hence NDE methods should be selected to address specific requirements.

This appendix surveys selected NDE methods that have been, or potentially could be, applied to coated structure. Several promising, emerging methods are also discussed. All NDE methods use some external source to produce a response from the sample that can be detected without causing a permanent change in the specimen. The information extracted by the measurement is determined both by the initial source/specimen interaction and the detection method. NDE methods form families of techniques classified by either source-detection method or source/specimen interaction. Table 6-1 lists some of the primary NDE methods that have been used for coating inspection grouped primarily by detection method.

OPTICAL METHODS

Optical methods in this discussion refer to ultraviolet, visual, and infrared spectroscopy and imaging. Optical methods measure light transmitted or reflected from a specimen as a function of wavelength or position on the sample for selected polarizations or angle of incidence. From these measurements, absorption or reflection spectra can be determined if the form of the incident light wave is known or, alternatively, if images can be formed. Since many coatings scatter light from the voids or interfaces between grains, absorption measurements made by these techniques involve contributions from intrinsic absorption and scattering.

In the ultraviolet and visible light case, the incident light does not penetrate far into the sample but is confined near the coating surface by scattering. While surface cracks and other features that have a surface expression can be detected, little information can be obtained about the coating/substrate interface. Scattering is wavelength-dependent, however, and decreases with increasing wavelength. As a result, scattering is low enough in the near-infrared that it is possible to observe reflection from the substrate and the interface between the substrate and coating and to identify regions where the coating (e.g., yttria-stabilized zirconia TBCs) has disbonded from the superalloy substrate.

Emission spectra can be obtained in the infrared region for heated samples and the optical properties of TBC materials, including the absorption coefficient, determined using Kirchoffs relations between emissivity, reflectivity, and absorptivity. Thermally stimulated emission spectra are much more difficult to obtain in the ultraviolet to visible ranges as seen from Planck's radiation law,

$$Q(\lambda) = C_1 (\lambda^{-5}) \{\exp(C_2/\lambda T) - \lambda\}^{-1}$$

where Q is radiative flux per unit wave length, λ is the wavelength, T the temperature, and C_1 and C_2 are the first and second radiation constants. For the temperature range of interest to turbine coatings applications, virtually all of the radiation is at wavelengths longer than 1 to 1.5 µm independent of the emissivity of the specimen. For yttria-zirconia and some other ceramics, the emissivity has been determined and the absorptivity derived (Thomas and Joseph, 1987; Thomas et al., 1988; Thomas, 1989). There is a relatively low intrinsic absorption in the 2- to 4-µm region, and this allows the substrate to be viewed directly if scattering is excessive. This region of transparency should be explored for other candidate TBC ceramics to assess the prospect of inspection of the interface with the substrate as well as the coating microstructure.

A second aspect of this region of high transparency is the role of radiative transport in heat transfer across a TBC. Some thermal-imaging methods described later in this appendix can determine the rate of heat transport by conduction. However, the relative importance of conductive versus radiative transport does not seem to be well understood and should be studied (see appendix B for a discussion of energy transport in these coatings). If the gas-emission spectra follow a black-body law at the operating temperatures of existing and proposed engine temperatures, the peak emissions fall within the region of coating transparency suggesting efficient radiative transport. If the radiative load is high, then strategies to increase the thermal efficiency of the TBC must minimize both conduction and radiation.

Visual inspection is widely used for NDE of coatings. For metallic coatings, the primary goals are to determine if the coating is spalled from the substrate and if either the coating or the substrate is cracked. The major failure mechanisms of loss of oxidation protection by loss of the protective metal-oxide layer or by interdiffusion of coating and substrate can lead to the attack of the coating/substrate interface and to cracking under thermal cycling. Spallation and cracking of the coating can be identified by visual inspection in some cases. While visual inspection is not quantitative and cannot determine the useful remaining life of the coating, it can assess current damage at some level.

Enhanced visual inspection methods may improve the quantitative aspects of the method and should be applied to in-service inspection. Methods based on computer-aided image processing should be standard methods of inspection of engine components for sizing of parts and location and quantification of surface defects, such as cracks and spalls. Sizing of individual surface features may provide statistical data on qualitative changes that would aid routine inspection. The changes of surface features with service time should be monitored to document the initial condition and the rate of change with time with the goal of improving lifetime prediction. Technical aids such as coherent fiber optics in the visible and infrared might allow visual inspection of some parts to be carried out in situ.

THERMAL METHODS

Thermal imaging methods use a source of energy to create a temperature field in a specimen and a detection method to monitor the temperature changes. The information obtained depends on the initial source distribution, the subsequent thermal diffusion, and the detection method used. The dependence on the detection method exists because detection is based on measuring a change in some physical property of the specimen caused by temperature. Different properties may have different temporal and spatial scales as discussed below.

Optical-Source Infrared Radiometric Detection

For optically opaque materials, amplitude-modulated (pulse or step) optical heating acts as a surface heat source equal to a flux of heat into the specimen surface. This flux induces a change in surface temperature with time based on the magnitude of the flux and the rate of thermal diffusion into the specimen.

For a layered specimen, the surface temperature initially follows the functional form of the uniform specimen but then deviates in a direction that depends on the characteristics of the second layer, the interface properties, and the depth of the second layer in the sample. The time at which the deviation occurs is the thermal transit time and can be used to determine the coating thickness if the thermal diffusivity is known, or alternatively, the thermal diffusivity if the thickness is known. For example, accuracy of measurement for the thickness of 200-μm TBC coatings is better than 5 percent.

This method can also confirm that an infrared image represents a disbonded region. The time-resolved thermal image quantitatively measures the extent of the disbond, whereas the passive infrared gives rapid views of the area. These methods may allow characterization of the coating/bondcoat/substrate interfaces as a function of time if the spatial resolution for imaging of the interface is adequate. The results suggest that infrared source and thermal detection methods can be complementary methods for monitoring the condition of ceramic TBCs on high-temperature substrates.

Other-Source Infrared Radiometric Detection

Other sources (i.e., non-optical) that heat internal regions of the coating or heat the substrate or bondcoat may provide the spatial resolution needed to detect substrate cracks or changes in conductivity by thermal cycling. An illustration is the localized induction heating (Lehtiniemi et al., 1991, 1993) of a plasma-spray-deposited ceramic on a high-temperature metal alloy. The inductive source heats the component resistively by currents induced in the substrate. Since cracks and other thermal features inhibit current flow and heat flow, this method can image subsurface features through the analysis of surface temperature distribution.

Microwave-source methods also have promise for TBC characterization. Many ceramic materials are relatively transparent to microwaves within the 10 GHz region. Hence microwaves can penetrate the coating with little loss and interact with specific features or constituents in the coating that may be present as contaminants that were deposited during the period of service. The resultant temperature changes may be detectable using infrared imaging methods. Thermally detected microwave spectroscopy has already been shown to have high spatial resolution (10 μm) and the ability to determine the presence of selected impurities in dielectric materials. As a speculative idea, perhaps the presence, or even the concentration, of sulfur or other fuel-related species could be determined at the coating/substrate interface by this method.

Radiometric Methods in Coating Deposition

Radiometric methods are illustrative of the use of NDE methods in process control applications (Moreau et al., 1991, 1992). In plasma spraying, the heat source is ceramic particles that are accelerated toward a substrate through a hot plasma flame. The particles melt in passing through the flame. The infrared emission from the particles is monitored while in passage from flame to substrate and on the substrate itself. Using fast infrared detection methods, time-temperature profiles of individual particles on the deposited surface can be measured as they cool. The heat transfer rate depends on the thermal conductivity of the deposited layer and the characteristics of the particle, including particle size. Issues affected by the rate of cooling are lateral spread on the coating, extent of crystallization, and size of crystallites. Some changes in overall morphology can be followed. This method might be of value in providing sensor input to allow real-time control of plasma spraying so that the desired microstructure is deposited.

Laser Acoustic Interferometric Imaging

There are a range of thermal imaging methods that use a thermal source to induce a mechanical response by thermal expansion. Images are formed by scanning an amplitude-modulated, focused laser beam over the sample and measuring the surface displacement of the specimen either at the point of heating or at some point close by. There are several factors that affect the response, including the peak height of the surface deformation produced by the heating and the lateral spread and hence the slope of the deformation. Such methods can provide information about the spatially diverse mechanical responses of a specimen. These methods may have utility in imaging spatial changes in the stiffness of interfacial bonds.

Induced Current Methods

Induced current methods use relatively low-frequency magnetic and electric fields of varied frequency as the excitation source. In the case of a magnetic source, the induced magnetic fields generated by the interaction of the source with a conducting sample are measured. This interaction takes the form of eddy currents, which are dispersive waves penetrating the specimen to a depth called the skin depth. This depth, d, can be expressed as $d = \{s\mu/\omega\}^{0.5}$ where s is the specimen conductivity, μ the permeability, and ω is the frequency. As in the case of thermal imaging, the depth sampling provided by the skin depth, and especially its frequency dependence, allows buried regions of a specimen to be examined in a single-sided measurement. Eddy-current methods can determine the electrical properties of individual layers in multilayer conducting materials and the properties of conducting regions below a dielectric layer. They can also determine the presence of a crack in a buried conductive layer and, in some cases, the crack density. From an NDE perspective, this could be of value in assessing changes in the bondcoat or substrate with time because of crack formation or interdiffusion.

A derivative of the eddy-current method called photoinductive imaging (Moulder et al., 1992) offers higher spatial resolution (micron scale) for many materials. Eddy-current methods can also measure coating thickness with good accuracy (Smith and Stephan, 1990).

Other electromagnetic methods related to eddy currents use a variety of magnetometers and magnetometer arrays to monitor specimen condition (Goldfine et al., 1993; Wickswo et al., 1993). These advantages derive from the ability of nonsearch coil methods to operate at low frequencies to spatially pattern the fields in the specimen (Goldfine et al., 1993).

Injected Current Methods

Injected current methods monitor the distribution of current injected into a material or component using sensitive magnetometer arrays. These arrays can be physical arrays, in which measurements are made in parallel, or scanned arrays. The injected currents can be of any frequency that can be detected, and hence the depth of penetration into the specimen is controlled. One goal of these methods is to monitor through-thickness changes in current distribution that result from spatial variations in conductivity that are either intrinsic (e.g., caused by deformation) or extrinsic (e.g., caused by cracks). Initial experiments suggest that these methods may have a place with further development.

Conventional Ultrasonic Methods

Ultrasonic methods monitor the amplitude, phase, and velocity of an ultrasonic wave injected into the sample. Again, both transmission and reflection modes are available. For a layered medium, such as the coating/substrate system, in addition to a bulk compressional and two shear waves, Rayleigh waves and other interfacial waves exist. The characteristics of these surface and interface waves depend on the thickness of the coating and the stiffness of the mechanical bond between the coating and the substrate. These methods offer some promise for special applications, especially if noncontact generation and detection methods at elevated temperatures are used.

Laser Ultrasonics and Interferometric Detection

This method is related to the laser acoustic interferometric imaging discussed previously but differs in that it uses the analysis techniques of ultrasonics. A pulsed laser source is used to produce an ultrasonic wave in the specimen, and the amplitude, phase, and arrival time of the wave are measured in a typical manner. The previous work could be viewed as the quasi-static limit of laser source ultrasonics.

X-ray Methods

X-ray methods based on radiography or tomography have clear applications to inspection of coated components. Examples include visualization of internal passages in turbine blades, the location of casting defects, and a determination of the crystallographic structure of substrates, especially directionally solidified and single-crystal materials. Synchotron-source X-ray topography has the advantage of allowing the crystal orientation and structure of an entire part to be visualized in a single measurement. While these methods require special facilities, they might be useful in process development.

Other methods are based on X-ray fluorescence and scattering (Verbinski, 1990) and are directed at the inspection of the coating. The X-ray scattering cross section is energy-dependent with elastic scattering (Compton scattering) dominating for low-atomic-number targets and medium energies and fluorescence dominating for high-atomic-number targets and low energies. In the case of fluorescence, energy-selective detection methods are used to provide some discrimination of elemental composition. Because both coating and substrate are based on high-atomic-number material, this method appears to have some promise for composition and coating density measurements.

There is a possibility that this method could be developed to include profiling of composition with depth and to monitor changes associated with exposure to the engine environment. In practice, however, the results have been of limited value because of counting-rate problems and difficulties in deconvoluting fluorescence line profiles.

EMERGING NDE METHODS FOR INSPECTING COATING SYSTEMS

X-ray Thermal Imaging: Composition with Depth

Another X-ray method that has only been applied to subsurface characterization of 10-μm multilayer films uses thermal detection to monitor X-ray absorption as a function of photon energy (Masujima et al., 1987). The energy absorbed from an X-ray source of variable energy increases near the characteristic wavelength of specific elements. This increased heat flux into the sample changes the volume and surface temperature of the sample. The temperature changes can be detected by methods that monitor either the surface temperature (e.g., radiometry) or the acoustic response of the sample as discussed earlier in this appendix. Depth profiling is possible based on the arrival time for thermal diffusion from the layer to the surface or on the acoustic propagation time to the detector. This approach has not been tried for coating systems, but it has been used to characterize subsurface nickel and copper multilayers in semiconductor systems. Its potential advantages are that it is single-sided, multi-element-sensitive, preferentially responsive to high-atomic-number elements, and able to depth profile. Its disadvantages are the need for a high flux, tunable X-ray source.

Transient Grating Characterization Methods

Transient grating characterization methods have been recently introduced for characterization of deposited layers on surfaces. In the simplest form, they involve formation of a linear array of lines of optical or other energy on the surface of the specimen using moire or interferometric methods. Detection is also optical through either light reflection or emission methods. Because the grating is both a thermal and an acoustic source, several detection modalities are possible. In the first case, thermal properties can be determined in the plane of the coating and across the coating in a single measurement. In the second case, ultrasonic waves of known spatial and temporal structure can be launched and reflections measured at very high rates. Thin coatings and multilayered systems can be analyzed even for submicron layers, such as those present in some of the nanostructured coatings being considered for TBC applications. The array allows the spatial frequency of the measurement to be varied and the spatial scale of characteristic defects to be evaluated. These methods have advantages in being noncontact, fast, and flexible in the information presented.

An example of possible applications of transient (picoseconds to microseconds) heating in coating characterization is the use of fast radiometric heating of TBC and NiAl coatings. In both cases, the time evolution of the temperature response under fast pulse heating was a relaxation whose time dependence is

$T(t) = t^{-a}$

where t is time and a < 0.5 (the value for one-dimensional diffusion). This result suggests that the thermal diffusion in these coatings is geometrically constrained by the coating structure. It may be possible to infer structure using these methods, perhaps even in the process of layer deposition.

Microwave Thermoreflectance

As discussed previously, many ceramics used for TBC applications are transparent to microwaves. Microwave reflectance measurements from the coating/substrate interface can be made to temperatures in excess of 1100°C and the reflectivity measured as a function of temperature. This raises the prospect of direct measurement of the substrate temperature through coatings. As another feature of the method, changes in reflectivity with melting and possibly with other phase changes in the material have been seen. This method is in the early phases of development but appears to have promise for application in all four areas of NDE applications.

Modulated Stress Methods

Transient grating methods can also be produced by modulated stresses in the plane of the specimen (Lesniak and Boyce, 1993). These methods are at much slower time scales than the interferometric methods but provide information about the response in the quasi-static regime for both thermal diffusion and stress relaxation. They have proven to be of value in other materials and coating systems and should be considered for high-temperature coatings.

REFERENCES

Goldfine, N.J., A.P. Washabaugh, S.V. Dearlove, and P.A. Guggenberg. 1993. Imposed w-k magnetometer and dielectrometer. Review of Progress in Quantitative Nondestructive Evaluation 12:1115-1122.

Lehtiniemi, R., J. Hartikainen, J. Rantala, J. Varis, and M. Luukkala. 1991. P. 441 in Review of Progress in Quantitative NDE, Vol. 11A, D.O Thompson and D.E. Chimenti, eds. New York: Plenum.

Lehtiniemi, R., J. Hartikainen, J. Rantala, J. Varis, and M. Luukkala. 1993. P. 1931 in Review of Progress in Quantitative NDE, Vol. 12B, D.O Thompson and D.E. Chimenti eds. New York: Plenum.

Lesniak, J., and B.R. Boyce. 1993. Forced diffusion thermography. Pp. 92-102 in the Proceedings of SPIE-The International Society for Optical Engineering Conference on Nondestructive Inspection of Aging Aircraft, Vol. 2001, San Diego, California. Bellingham, Washington: SPIE.

Masujima, T., H. Kawata, Y. Amemiya, N. Kamiya, T. Katsura, T. Iwamoto, H. Yoshida, H. Imai, and M. Ando. 1987. X-ray photoacoustic effect of solid materials. Chemical Letters 973-976.

Moreau, C., P. Cielo, M. Lamontagne, S. Dallaire, J.C. Krapez, and M. Vardelle. 1991. Temperature evolution of plasma sprayed niobium particles impacting on a substrate. Surface and Coatings Technology 46:173-187.

Moreau, C., M. Lamontagne, and P. Cielo. 1992. Influence of coating thickness of the cooling rates of plasma sprayed particles impinging on a substrate. Surface and Coatings Technology 53:107-114.

Moulder, J.C., E. Uzal, and J.H. Rose. 1992. Thickness and conductivity of layers from eddy current measurements. Review of Science Instruction 63:3455-3465.

Smith, K., and R. Stephan. 1990. Protective Coatings Nondestructive Evaluation. Final Report. Contract F33615-87C-5221. Wright-Patterson Air Force Base, Ohio: U.S. Air Force Wright Research and Development Center .

Thomas, M.E. 1989. A computer code for modeling optical properties of window materials. Proceedings of SPIE: Window and Dome Technologies and Materials 1112:260-267.

Thomas, M.E., and R.I. Joseph. 1987. Characterization of the complex index of refraction for sapphire, spinel, alon and yttria. Paper presented at the Meeting of the IRIS Specialty Group on Infrared Materials, June 9-10.

Thomas, M.E., R.I. Joseph, and W.J. Tropf. 1988. Infrared transmission properties of sapphire, spinel, yttria, and ALON as a function of temperature and frequency. Applied Optics 27:239-245.

Verbinski, V.V. 1990. Protective Coatings Nondestructive Evaluation. Final Report. Contract F33615-87-C-5221. Wright-Patterson Air Force Base, Ohio.: U.S. Air Force Wright Research and Development Center.

Wickswo, J., D.C. Hurley, Y.P. Ma, S. Tan, and J.P. Wikswo, Jr. 1993. P. 633 in Review of Progress in Quantitative NDE, Vol. 12A, D.O. Thompson and D.E. Chimenti, eds. New York: Plenum.

Appendix D

Modeling of Coating Degradation

The availability of accurate models for predicting the life of coatings will become essential as future engines become increasingly reliant on coatings to protect the components from the higher firing temperatures required to improve performance. As discussed in chapter 4, coating life at high temperature is determined by many degradation mechanisms, primarily oxidation, hot corrosion, and fatigue.

CURRENT STATUS

Models for life prediction based on high-temperature oxidation and hot corrosion of coatings are not well developed and are largely empirical or semi-empirical. Existing approaches test coatings in furnaces or burner rigs to determine life and to use experience factors to relate the life in the furnace or burner rig to the life in the engine. These approaches are unsatisfactory but are used in the absence of better methods and in the presence of a large, even if questionable, database of test results and correlations. Problems with these approaches include the following:

- The approach does not enable the life of a coating in a new engine design to be accurately determined but only to be estimated.
- Burner-rig tests typically exhibit a large amount of scatter and are often useful only for comparison with other laboratory reference materials.
- Furnace testing does not simulate the high-velocity gases and high heat fluxes to which coatings are exposed and which can be important to their behavior.
- Hot-corrosion life depends on the deposition of the corrodants on the surface. The rate of this deposition is not given by the burner rig or other methods used to test hot-corrosion resistance.
- The actual environments, including the erosive species, to which the coatings are exposed vary widely, especially for industrial engines. These conditions are difficult, if not impossible, to simulate in the laboratory and are not always known.

Although not satisfactory, determination of high-temperature oxidation life by these methods can be done for existing engines after sufficient experience has been accumulated on the engine. However, hot-corrosion life requires a determination of the deposition of the contaminants on the surface, which depends on the environment and the flow profiles through the engine.

Modeling of high-temperature oxidation life has been developed by Probst and Lowell (1988), based on isothermal oxidation and thermal cycling.

$$\xi, W_{ox} \text{ or } \frac{\Delta m}{A} \propto \sqrt{k_p t}$$

where ξ is the scale thickness, W_{ox} is the weight of the oxide scale, $\Delta m/A$ is the mass gain per unit area, k_p is the parabolic rate constant, and t is time. k_p is related to temperature by an Arrhenius rate law:

$$k_p = k_{po} \exp\left(\frac{-Q}{RT}\right)$$

where k_{po} is a material constant, Q is the activation energy, R is the universal gas constant, and T is the absolute temperature in K or °R. k_{po} is a function of oxygen partial pressure, but it is not a strong function. The model tracks the formation of the oxide by a parabolic growth law (although other growth kinetics can be used), and the spallation of the oxide due to thermal cycling.

$$W_{spall} = Q_0 W^2_{oxide}$$

where W_{spall} is the weight per unit area of the oxide spalled, Q_0 is an experimentally derived material constant, and W_{oxide} is the weight per unit area of the oxide prior to spalling. (Only a portion of the oxide may spall and some of the oxide may remain adherent to the coating.) The spallation equations are empirical and are based on careful experimental measurement of the weight of the spalled oxide. Barrett and Lowell (1975) have also developed a mathematical expression to fit cyclic oxidation data. Both of these approaches rely heavily on empirical methods. The model for thermal cycling is attractive but is based on experimental calibrations of the amount of oxide spalled instead of a more fundamental, predictive model. Furthermore, spallation data for only one

APPENDIX D

chromia-forming system and one alumina-forming system are available (at least in the public domain). The committee is not aware of any actual uses of these models.

Wright et al. (1991) have developed models for erosion modified oxidation, based on an erosion rate assumed to increase with the thickness of the oxide.

$$\xi = act - \frac{\xi_1 (1 - \xi_1^k)}{k + 1}$$

where ξ is the scale thickness as a function of time, t is the time between erosion events, a is an oxidation rate term, c is the Pilling-Bedworth ratio,[1] ξ_1 is the scale thickness after the last erosion event ($\xi_1 = 0$ at t = 0), and k is a term describing the oxidation kinetics (k = 0 for linear kinetics and k = 1 for parabolic kinetics). A numerical solution method is used to determine the degradation of the coating.

Chang et al. (1990) have studied the interaction of erosion and oxidation and have defined the various regimes of erosion and oxidation. These last two models were developed for coal gasification and fluidized bed applications. Bernstein (unpublished work) has combined the oxidation, spallation, and erosion models for industrial gas turbines and has added erosion rate terms. An example of the approach to modeling based on correlation with engine service experience is given by Strangman (1990). For high-temperature oxidation, Strangman computes the oxidation rate as the product of an engine experience factor, a burner-rig calibration constant, and a term for the temperature dependence of oxidation. For hot corrosion, Strangman uses a lengthy expression as follows:

Hot-Corrosion Rate = Salt Deposition Factor · Salt Corrosivity Factor Salt Deposition Factor = {Salt Vapor Deposition + (Component Experience · Particulate Salt Deposition) + Fuel Salt Deposition ·
Sulfate Formation + Fuel Vanadium Deposition + Salt RetentionEvaporation}
Salt Corrosivity Factor = {(Engine Component Experience · Burner-Rig Calibration Factor · Acidic Temperature Factor · Acidic Sulfur Factor) + (Engine Component Experience · Burner-Rig Calibration Factor · Basic Temperature Factor · Basic Sulfur Factor) + (Engine Component Experience . Burner-Rig Calibration Factor · Vanadate Sulfur Factor) } · Fraction Salt Molten.

The total environmental attack is the sum of the hot-corrosion rate and the oxidation rate. These equations are solved by means of a computer program that has additional calculations for many of the terms shown above.

Nesbitt (1984, 1989) has developed a diffusion-based approach to determine the oxidation life of coatings and bulk NiAl. For bulk materials, the model computes the flux of aluminum to the surface and the removal of aluminum by oxidation. When the aluminum flux becomes less than the aluminum consumption by formation of Al_2O_3, $NiAl_2O_4$ becomes the stable oxide. Since this spinel is not protective, the bulk metal will be attacked at this point. For coatings, the flux of aluminum in the coating to the surface is computed and the interdiffusion of the coating with the base metal is also computed. This interdiffusion reduces the aluminum concentration (actually aluminum activity) in the coating, shortening the life of the coating.

Fracture mechanics approaches to modeling coating spallation have been and are continuing to be developed (Evans, 1994). They usually predict a dependence on the square of the temperature change, ΔT

$$W_{spall} \propto (\Delta T)^2$$

However, the utility of these models to predict coating degradation, as well as their agreement with experimental data, remain to be determined. Continuing work on modeling coating degradation by fracture mechanics may or may not prove useful. Until a better understanding of the physical mechanisms of coating degradation and the loss of the protective oxide is obtained, it is not possible to say if this is a fracture problem or not. It must be remembered that the oxide is in a dynamic flux, in which its composition, morphology, and thickness are changing as a function of time, and its properties depend on the location within the oxide scale.

The committee did not seek, nor obtain, details of proprietary models used by engine manufacturers. However, the models used by the manufacturers, as exemplified by Strangman's model, usually are empirical in nature.

FUTURE DIRECTIONS

Thermal Barrier Coating Modeling

TBC modeling is important for life prediction, understanding TBC behavior, and design of new coatings. Life prediction is the primary modeling issue that is needed to obtain the maximum benefit of TBCs apart from process modeling, which is covered elsewhere in this report. In spite of the use of TBCs for over 30 years, there has been relatively little activity in modeling their durability and reliability. In the past 10 to 15 years, there have been several large studies sponsored by NASA. Modeling is also necessary for more efficient design of new TBCs and for examination of the complex TBC failure mechanisms. This subsection discusses previous and current models of the life and durability of TBCs and directions for future development.

It should be noted that most of the modeling has been for high-temperature oxidation combined with thermal cycling. Hot corrosion of TBCs is not well understood, and there has

---

[1] The Pilling-Bedworth ratio is the ratio of the volume of the oxide formed to the volume of the metal consumed

APPENDIX D

been only a limited effort in modeling the hot-corrosion life of TBCs.

Early TBC studies combined modeling and experimental work to define some of the important factors in TBC durability (Sevcik and Stoner, 1978; Cassenti et al., 1981), which are:

- the effect of residual stress generated during processing on coating spallation
- the correlation of in-plane compressive stresses on ceramic layer spallation
- the importance of maximum operating temperatures on TBC life

An oxidation-based failure model was proposed by Miller (1984) assuming that the strains imposed on the ceramic layer were caused by thermal-cycle effects and that time at temperature effects acted to increase the effective thermal-cycle strains caused by oxidation.

$$N_f = \left(\frac{\varepsilon_f}{\varepsilon_r}\right)^{-b}$$

$$\varepsilon_r = (\varepsilon_f - \varepsilon_c)\left(\frac{W}{W_c}\right)^m + \varepsilon_c$$

where N_f is the number of cycles to failure, ε_f is the failure strain in one cycle with no oxidation, ε_r is the thermal expansion mismatch strain, W is the weight gain by oxidation, and W_c is the weight gain that should cause failure in one cycle. b and m are material constants determined from the experimental data. Miller's model correlated the cyclic life of a specific coating system tested at 1100°C for different thermal cycle times.

In the 1980s, significant advances in TBC modeling capability occurred for applications of specific TBCs. DeMasi et al. (1989) used a model similar to Miller's for plasma-sprayed TBCs, except that the thermal-cycle strains were included using a finite-element analysis that incorporated a Walker constitutive model for inelastic deformation of the ceramic during thermal cycling, and the strain to failure in one cycle, ε_f, changes with oxidation.

$$N_f = A\left(\frac{\Delta\varepsilon_c}{\varepsilon_f}\right)^{-b}$$

$$\varepsilon_f = \varepsilon_c\left(1 - \frac{W}{W_c}\right) + \Delta\varepsilon_p\left(\frac{W}{W_c}\right)$$

where $\Delta\varepsilon_p$ is the inelastic strain range, ε_c is ε_f at W = 0, and the other terms are as before. This same model was used for life predictions of EB-PVD (electron-beam physical vapor deposition) coatings (Manning-Meier et al., 1991), except that the EB-PVD zirconia ceramic layer was assumed elastic, the bondcoat was assumed to have inelastic behavior, and the thermal strain of the alumina scale was included. Therefore, the important cyclic strain for the EB-PVD ceramic was between the alumina layer and the substrate. Both models showed reasonable correlation to the experimental results.

Hillery et al. (1988) focused on the edges of TBCs as failure initiation sites for some TBC applications. Edge failures were driven by shear stresses in addition to normal stresses and oxidation. Oxidation was included through changes in the time-dependent material behavior and the stress state. This model was also found to give reasonable agreement with experiment.

Strangman et al. (1987) used a linear damage rule (damage from thermal cycling, oxidation, and hot corrosion) to reduce the strain-to-failure of the ceramic (instead of increasing the effective strain per cycle). This model reduces to a MansonCoffin relation. Again, the agreement to the experimentally measured lives was reasonable.

For each case the intent was to develop a model that would predict life for a given coating system rather than to determine the active degradation mechanisms. To do this, each model incorporated information known about the active degradation mechanisms for the specific coating and application domain, which was for aircraft engines. In addition, the models were heavily dependent on calibration to experimental data to predict the correct absolute lifetimes.

The vast majority of TBC modeling, including those models cited above, has been for current aircraft applications, which have a much different operational profile than land-based or marine gas turbines. Thus, for nonaircraft applications, it may be expected that the relative contributions and importance of different mechanisms to TBC failure would be different. These mechanisms include oxidation of the bondcoat, top-coat creep, sintering of the top coat, and the mechanical behavior of the alumina scale. These mechanisms have not yet been adequately characterized experimentally or theoretically.

A few models address long-term oxidation behavior and loss of alumina-forming capability ("wear out") of overlay coatings (Nesbitt and Heckel, 1984) and bondcoats for TBCs (Lee and Sisson, 1994). In these models, the life of the TBC is determined by the life of the bondcoat. While these models are not appropriate for aircraft turbines, they may address one aspect of failure for land-based and marine-based turbines.

Comprehensive models that can predict TBC life from first principles are necessary to be able to account for the wide range of conditions to which TBCs are exposed. Before a model can be developed to predict TBC life from first principles, a large number of factors must first be investigated. This is because many of the potential contributors to TBC failure have not been adequately characterized because of the extreme difficulty in experimentally isolating these factors. Understanding of TBC behavior has been enhanced by computer modeling.

Computer modeling has been used to investigate TBC behavior by isolating and investigating issues not easily amenable to experimental study. Chang et al. (1987) and Phucharoen (1990) developed a finite-element model that examined the effect of the rough interface between the bondcoat and ceramic layer, the mechanical effects of oxidation on the stresses in ceramic layers, and the effect of bondcoat properties on TBC life. The most important of these issues are the effect of interface geometry and oxidation on TBC stress levels. Another recent computer model examining failure mechanisms, instead of life prediction, used the observed bondcoat creep response (Brindley and Whittenberger, 1993) to show that bondcoat creep can result in substantial increases in the ceramic layer delamination stresses (Brindley, 1995), which should result in decreased TBC cyclic lives. Ferguson et al. (1994) included bondcoat and top-coat creep along with a rough interface to provide a clearer picture of the role of these factors increasing the stresses generated in a TBC during thermal cycling. Apart from failure mechanisms, theoretical modeling may also shed light on heat transfer through TBCs, which in turn may guide design of more-insulating coatings.

The models discussed above have been used to guide TBC development and design. A more "design capable" model for TBCs is necessary to facilitate more rapid development and implementation of TBCs. Such a model would need to incorporate operating conditions, life prediction from first principles, and thermal modeling from first principles. It will also be necessary to establish a materials property database to support the model.

High-Temperature Oxidation

The current approach to modeling high-temperature oxidation is manageable for the original equipment manufacturer (OEM), although not desirable. It is based on laboratory testing and correlation with field experience, which has been developed over a number of years. One of the principal difficulties is the ability to accurately predict the surface temperature of the blading. Although not a coating issue, the difficulty of predicting surface metal temperature is a barrier that limits the ability of any coating degradation model to properly determine the life of the coating.

There is no current approach to predicting the degradation of the coating that the operator of the engine can use. The operator relies on OEM guidelines, which are based on design considerations and accumulated field experience. The engine operators need models that allow them to predict the coating life for their operation. This is particularly true for industrial engines that experience a wide range of conditions. Any models developed for the operator will also assist the OEM.

The current approaches to modeling do not provide the necessary insight that allow the development of more-oxidation-resistant coatings. Studies of a fundamental nature, as opposed to an engineering evaluation, are needed to determine how coatings are exposed and how they perform in the field, particularly for industrial gas turbines. Engine exposure can be fundamentally different than laboratory testing, and the nature of the protection by the coating will depend on the local exposure environment of the coating. Once these environments are determined and the coating response in the engine is established, laboratory studies simulating this behavior can be conducted to understand and predict the kinetics of the degradation of the coating. These studies can lead to the development of improved coatings and coating degradation models.

Hot Corrosion

A great deal is understood about the mechanisms and chemistry of hot corrosion, as discussed elsewhere in this report. The prediction of hot-corrosion attack of coatings in the field cannot be reliably performed, however. This inability is due in part to the difficulty in predicting the generation and deposition of corrodants and knowing the chemistry of the deposits and the surface metal temperatures.

As for high-temperature oxidation, the current approaches to modeling the life of a coating also do not provide the necessary insight that allow the development of more-corrosion-resistant coatings. (The mechanistic and chemical studies of hot corrosion do provide this insight, however.) Studies of a fundamental nature, as opposed to an engineering evaluation, are needed to determine the exposure environment of coatings and how they perform in the field, particularly for industrial gas turbines. As stated above, engine exposure can be fundamentally different from laboratory testing, and the nature of the protection of the coating will depend on the local exposure environment of the coating. Once these environments are determined and the coating response in the engine is established, laboratory studies simulating this behavior can be conducted to understand and predict the kinetics of the coating degradation. These studies can lead to the development of improved coatings and coating degradation models.

Modeling of hot corrosion is currently not as important as for high-temperature oxidation for industrial gas turbines because of the use of clean natural gas. Hot corrosion is still of importance to aircraft and marine engines, however. It will also be of importance to industrial engines if coal gasification or biomass fuels are used. Life prediction based on hot-corrosion attack is particularly difficult because the onset of attack may be caused by a local flaw in the protective oxide that permits contact of the fused salt film with the coating. Thus the phenomenon does not obey some steady-state predictable kinetics.

Life-Prediction Models

One important aspect of the immature status of life-prediction models deserves emphasis. The overall goal for life-prediction models is to enable the component designer or user of the engine to make successful decisions regarding the selection of the superalloy, coating, and coating thickness distribution. Design decisions regarding the superalloy and coating must be made early in the engine design process (i.e., when the airfoil is still on paper). The type of information that should be available to the designer when selecting the coating and superalloy to be used is engine design parameters (e.g., gas temperature, velocity, and pressure; substrate and coating temperatures; fuel/air ratio; inlet filtration efficiency; stress; strain), anticipated application usage (e.g., fuel, location, altitude, and times at duty-cycle power points), available coating and superalloy properties, and prior engine experience knowledge bases. To make successful decisions regarding the selection of the superalloy, coating, and coating thickness distribution, the component designer or user of the engine needs mechanistic life-prediction methods that are expressed in terms that the designer and user of the engine can recognize, control, and, in some cases, adjust. For instance, life-prediction methods are needed that will enable designers to avoid thermomechanical fatigue cracking of coated superalloy components. Environmentally enhanced thermomechanical fatigue is a special case of low-cycle fatigue that is not adequately predicted with isothermal low-cycle fatigue data and models.

RECOMMENDATIONS

The understanding of the degradation and failure mechanisms of high-temperature coatings in the field need to be improved, particularly with respect to the effects of engine operation and environment on the coating performance (e.g., thermal cycling).

Qualitative and quantitative models need to be developed that predict coating life based on mechanisms observed in operational engines. These models need to be applied to optimize coating development and to estimate more accurately the remaining coating life for in-service engines. However, property measurements using reliable test methods are required before modeling.

Methods to determine the deposition rate of corrodants on in-service hot-structure components need to be developed that can be monitored by the engine operators.

REFERENCES

Barrett, C.A., and C.E. Lowell. 1975. Oxidation of Metals 9:307-355.

Brindley, W.J. 1995. Properties of plasma sprayed bond coats. Pp. 189-202 in the Proceedings of the Workshop on Thermal Barrier Coating. NASA-CP-3312. Cleveland, Ohio: National Aeronautics and Space Administration Lewis Research Center.

Brindley, W.J., and J.D. Whittenberger. 1993. Stress relaxation of low pressure plasma-sprayed NiCrAlY alloys. Materials Science and Engineering A 163(1):33-41.

Cassenti, B.N., A.M. Brickley, and G.C. Sinko. 1981. Thermal and stress analysis of thermal barrier coatings. In AIAA/SAE/ASME 17th Joint Propulsion Conference, Colorado Springs, July 27-29. New York: American Institute of Aeronautics and Astronautics, Inc.

Chang, G.C., W. Phucharoen, R.A. Miller. 1987. Finite element thermal stress solutions for thermal barrier coatings. Surface Coating Technology 32:307-325.

Chang, S.L., F.S. Pettit, and N. Birks. 1990. Interaction between erosion and high-temperature corrosion of metals: the erosion-affected oxidation regime. Oxidation of Metals 34(1/2):23.

Evans, H.E. 1994. Modeling oxide spallation. Materials at High Temperatures 12(2-3):219-227.

DeMasi, J.T., K.D. Sheffler, and M. Ortiz. 1989. Thermal Barrier Coating Life Prediction Model Development-Phase I. Final Report. NASA-CR-182230. Washington, D.C.: National Aeronautics and Space Administration.

Ferguson, B.L., G.J. Petrus, and M. Ordillas. 1994. A Software Tool to Design Thermal Barrier Coatings. Final Report. NASA Contract NAS3-2728. Washington, D.C.: National Aeronautics and Space Administration.

Hillery, R.V., B.H. Pilsner, R.L. McKnight, T.S. Cook, and M.S. Hartle. 1988. Thermal Barrier Coating Life Prediction Model Development. Final Report. NASA-CR180807. Washington, D.C.: National Aeronautics and Space Administration.

Lee, E.Y., and R.D. Sisson. 1994. The effect of bond coat oxidation on the failure of thermal barrier coatings: thermal spray industrial applications. Pp. 55-59 in Proceedings of the 7th National Thermal Spray Conference, Boston, Mass., June 20-24, C.C. Berndt and S. Sampath, eds. Materials Park, Ohio: ASM International.

Manning-Meier, S., D.M. Nissley, K.D. Sheffler, and T.A. Cruse. 1991. Thermal Barrier Coating Life Prediction Model Development. ASME Paper 91-GT-40. New York: American Society of Mechanical Engineers.

Miller, R.A. 1984. Oxidation based model for thermal barrier coating life. Journal of the American Ceramic Society 67(8):517.

Nesbitt, J.A. 1989. Predicting minimum Al concentrations for protective scale formation on Ni-base alloys. Journal of the Electrochemical Society 136(5): 1511-1527.

Nesbitt, J.A., and R.W. Heckel. 1984. Modeling of degradation and failure of Ni-Cr-Al overlay coatings. Thin Solid Films 119:281-290.

Phucharoen, W. 1990. Unpublished Ph.D. dissertation, Cleveland State University.

Probst, H.B., and C.E. Lowell. 1988. Computer simulation of cyclic oxidation. Journal of Metals 40(10):18.

Sevcik, W.R., and B.L. Stoner. 1978. An Analytical Study of Thermal Barrier Coated First Stage Blades in a JT9D Engine. NASA-CR-13560. Washington, D.C.: National Aeronautics and Space Administration.

Strangman, T.E. 1990. Turbine coating life prediction model. In 1990 Proceedings of the Workshop on Coatings for Advanced Engines, August. Washington, D.C.: U.S. Department of Energy.

Strangman, T.E., A. Liu, and J. Neumann. 1987. Thermal Barrier Coating Life-Prediction Model Development. Final Report. NASA-CR-179648. Washington, D.C.: National Aeronautics and Space Administration.

Wright, I.G., V.K. Sethis, and V. Nagarajan. 1991. An approach to describing the simultaneous erosion and high-temperature oxidation of alloys. Journal of Engineering for Gas Turbines and Power 113(0ctober):616.

Appendix E

Manufacturing Technologies of Coating Processes

This appendix reviews various coating process technologies: diffusion coating, thermal spray (particulate deposition), physical vapor deposition (atomistic or molecular transfer), and sputtering (atomistic deposition). Electrospark deposition is also discussed. Auxiliary processes and quality control procedures are also reviewed.

DIFFUSION COATING PROCESSES

Pack Cementation

In the pack-cementation process, the parts to be coated are placed inside a vented or purged retort and embedded in a pack mixture consisting of an inert powder (e.g., aluminum oxide), a source (e.g., pure or pre-alloyed aluminum), and typically a halide activator salt to generate the transporting vapor species. The retort is placed in a furnace and brought to the coating temperature. A protective atmosphere (generally argon or hydrogen) contacts the pack powders to prevent their oxidation. Pack cementation is generally conducted at temperatures between 650 and 1090°C (1200 and 2000°F) between 2 and 20 hours. Aluminizing is by far the most common process, although packs have been used to transfer chromium, silicon, and hafnium (Goward and Cannon, 1988). Recently, zirconium and yttrium have also been reported to be codeposited with aluminum (Bianco and Rapp, 1993).

Variants of pack cementation include slurry-fusion and electrophoretic plating where the pack is either sprayed or electroplated onto the part to be coated. The components are then heat treated in a protective atmosphere to form the coating.

For the intended thickness and aluminum concentration, there are over 20 different production aluminizing processes in use. The principal differences are in the type of coating desired. The effective pack aluminum activity is selected through a choice of the source-powder composition (commonly alloys of aluminum and chromium or cobalt), the amount of the source powder compared to the inert oxide, the respective powder size distributions, and the halide activator used. The temperature used for forming the coating determines to a large extent the degree of outward nickel diffusion obtained during processing. At temperatures above -1050°C (1925°F), coatings with more outward growth, characteristic for packs with reduced thermodynamic aluminum activity, form NiAl with lower aluminum contents. Below -1000°C (1825°F), coatings with more inward growth produce NiAl with higher aluminum contents. The low-temperature *high-activity* coatings may require an additional thermal diffusion step *out-of-pack* to reduce the composition and property gradients (Goward and Boone, 1971).

Out-of-Pack Methods

For the above-the-pack process, the parts to be coated are fixtured out of contact with the pack mixture. The coating vapors are transported to the parts by an inert carrier gas or purge gas and, through the use of two packs, both internal and external surfaces can be separately coated (Levine and Caves, 1973; Benden and Parzuchowski, 1979; Bianco and Rapp, 1993). Other variations are also in use, such as pulse aluminizing and SNECMA vapor phase aluminizing (Gauje and Morbioli, 1982). As with pack cementation, for pulse aluminizing and SNECMA vapor phase aluminizing, the retorts are loaded into an appropriate furnace for the thermal cycle. Because of the increased path length for vapor transport, the out-of-pack methods favor an increased tendency for outward growth and a generally reduced deposition rate.

Chemical Vapor Deposition

In the chemical vapor deposition (CVD) process, the coating deposition reaction takes place at the part surface. The parts to be coated are fixtured in a retort and placed in the CVD furnace. The reaction gases are metered into the reactor from external sources. Separate sources generally supply the internal and external airfoil coating circuits. Typical precursor gases for CVD aluminizing are HCl or HF, and these are passed over a source of aluminum under specific conditions of temperature, pressure, and flow rate to generate gaseous aluminum chloride or fluoride compounds. By proper selection of processing conditions, the CVD process can be varied to create the gas phase analogous to the pack-cementation coatings. The production of the platinum-modified aluminide on the external airfoil surfaces with a simple aluminide coating on the internal passages is currently of commercial interest (Smith and Boone, 1990).

THERMAL SPRAY PROCESSES

Plasma Spraying

A plasma gun functions through the operation of a stable nontransferred direct-current electric arc between a watercooled thoriated tungsten cathode and an annular watercooled copper anode. An inert plasma gas, which is generally argon with a few percent of an enthalpy-enhancing gas (e.g., hydrogen), is introduced as a vortex within the interior of the gun. The electric arc between the cathode and the anode creates a plasma arc within this gas. This ionized gas exits from the nozzle, where ionic recombination occurs, releasing enthalpy and yielding an effective temperature of the order of 15,000 K for the typical torch operating at 40 kW. The plasma temperature drops off rapidly from the exit of the anode. The feedstock powder is injected either internally or externally into the exiting plasma flame. The powder particles, approximately 40 microns in diameter, are accelerated and melted in the flame on their path to the target substrate, where they impact and undergo rapid cooling (10^6 K/sec) and solidification. The particle velocity can range from 100 to several hundred meters per second depending on spray parameters and the ambient atmosphere (e.g., low-pressure plasma spray; see below).

Plasma spray is generally used to form deposits of greater than 50-microns thickness of numerous industrial materials, including nickel-base and ferrous alloys and refractory ceramics (e.g., aluminum oxide and zirconia based ceramics). To approach theoretical bulk density and extremely high adhesion strength for high-performance applications, the plasma spray of metallic coatings is carried out in a reduced-pressure inert gas chamber. This *vacuum plasma* or low-pressure plasma spray (LPPS) operates at pressures of between 50 and 200 mbar. Shrouded flames can also be used (e.g., as developed by Praxair), where argon or nitrogen excludes oxygen from the vicinity of the flame and the work piece.

Although traditional plasma spray guns are gas-vortex-stabilized and operate in the 40- to 80-kW power range, it is possible to operate at considerably higher power levels (i.e., in the range of 160 kW and beyond) by using water stabilization. With a material throughput of about 30 times that of gas-stabilized torches, these high-production rates allow the manufacture of thick TBCs, such as those required in abradable seal applications (Chraska and Hrabovsky, 1992).

The key features of plasma spraying include the following:

- deposition of metals, ceramics, or any combinations of these materials
- formation of microstructures with fine, noncolumnar, equiaxed grains
- ability to produce homogeneous coatings that do not change in composition with thickness (length of deposition time)
- ability to change from depositing a metal to a continuously varying mixture of metal and ceramic (i.e., functionally graded materials)
- ability to achieve high deposition rates (>4 kg/hr)
- ability to process materials in virtually any environment (e.g., air, reduced-pressure inert gas, high pressure, under water; see the following section)

New and improved powder-processing methods have led to powders having predictable and controllable compositions and well-delineated particle-size distributions, an important parameter in the plasma-spray process. Ceramic powders for plasma spray are processed in diverse ways. Both particle size and shape are important controlling variables. In particular, the particle-size distribution has a great influence on the velocity and melt behavior in the plasma flame. These issues are discussed extensively in the literature (Herman, 1991). The deposition efficiency (i.e., the percentage of powder that actually becomes part of the target body) is of obvious economical importance and arguably represents a measure of deposit *quality*. Extensive literature exists on feedstock alloys and ceramic materials for plasma spray.

In addition to the starting material and its particle-size distribution, the microstructure of a plasma-sprayed coating also depends on the processing parameters, including plasma power, plasma gas composition, pressures and flow rates, powder injection details and carrier flow, torch/substrate distance, as well as other subtle factors. These parameters are sometimes interconnected in complex ways, leading to cross-terms in the process of parameterization. Statistical process control analysis is used extensively in thermal spray technology. This subject has been covered extensively in papers in the annual proceedings of the National Thermal Spray Conference (ASM International). Process control is becoming more common in plasma-spray processing of high-performance coatings. A clear goal is to achieve on-line feedback control of the process (i.e., intelligent processing of materials), which requires a much more detailed understanding of the process.

Low-Pressure Plasma Spray

The LPPS process was developed by Muehlberger (1973) in the early 1970s and gained widespread commercial use in the mid-1980s. It is competitive with electron-beam physical vapor deposition (EB-PVD) for the production of high-quality metallic (MCrAlY) coatings for certain applications because of the compositional flexibility afforded and the high coating rates achieved through molten droplet transfer.

As with all thermal spray processes, LPPS is limited to line-of-sight deposition. Individual parts are fixtured on a part manipulator inside a load-locked transfer chamber. The load lock is pumped down and the parts are preheated to about 900-1000°C before being transferred to the coating chamber. The plasma guns currently in use are in the 50- to 120-kW range and generally use argon-helium or argon-hydrogen gas mixtures to generate the plasma jet. Prior to initiation of the powder feed, the part is usually treated through reverse transferred arc sputtering to remove any traces of oxide that may have formed during preheat. The part is then plasma sprayed in a nontransferred mode. The coating distribution is controlled by the motions of the computer-controlled gun and the part. Typical parameters for turbine blade coating include a gun-to-substrate distance of ~10 to 16 in. at a chamber pressure of 30-60 Torr and a gun power of 80 kW. Powder feed rates vary from ~3 to 20 kg/hr, depending on the application.

The LPPS process has found its greatest use for large-power-generation turbine buckets where the coating is generally applied in the 7to 15-mil range and in turbo-fan-blade applications in the 4- to 6-mil range.

High-Velocity Oxy-Fuel Processes

High-velocity oxy-fuel (HVOF) techniques have proven capable of depositing a wide range of hard facings, metals, and cermets, and have demonstrated excellent deposit integrity and density (Kaufold et al., 1990). These high-velocity combustion torches are currently competing with both LPPS and detonation gun for the production of high-performance metallic aircraft coatings. Thus, HVOF can effectively apply bondcoats (Russo and Dorfman, 1995). More recently developed versions of this class of guns use oxidizers other than oxygen, for example, high-velocity air fuel (HVAF). HVAF guns offer both safer and more economical operation than HVOF and are being considered for aircraft engine applications.

PHYSICAL VAPOR DEPOSITION PROCESSES

The physical vapor deposition (PVD) process can deposit coatings of metal, alloys, and ceramics on most materials and on a wide range of shapes. Since this is a process limited to application by line-of-sight, complete coating coverage is achieved by manipulating the part during the coating cycle with a complex mechanical system.

Electron-beam guns are favored for supplying the energy necessary for evaporation because of their ability to achieve very high-energy densities compared with other methods of heating. The material to be evaporated is fed into the chamber through a water-cooled copper crucible where a rapidly scanning electron beam(s) melts the surface causing a vapor cloud to form above the ingot. Constant pool levels are maintained through the use of continuous ingot feed and pool-height sensors. Commercial TBCs have been successfully produced by EB-PVD. For yttria-stabilized zirconia deposition, an oxygen bleed is introduced into the vapor cloud to maintain coating stoichiometry. Because of the high melting point of yttria-stabilized zirconia (~3000°C), very shallow liquid pools are the norm, and temperatures at the point of beam impact are on the order of 5000°C. In PVD coatings, the structure has been found to be a strong function of the ratio of the substrate temperature to the melting point of the material being deposited. To achieve the desired columnar ceramic coating structure, the parts must be preheated to ~1000°C. This is typically accomplished by preheat chambers built into the load locks and through the use of over-source heaters in the main coating chamber. For increased filament life, the electron-beam guns are generally differentially pumped.

While the basic process of evaporation is relatively simple, there are significant differences in the equipment design for the production of turbine blade coaters. The most significant features that make up these differences are (1) electron-beam gun design; (2) number and location of electron-beam guns; (3) capability of the part manipulator; (4) number of load locks and part manipulators; (5) overall layout of the guns, part manipulators, and evaporation sources; and (6) overall level of complexity and automation of process variables.

SPUTTERING

Sputtering, an alternate vacuum process, relies on a reduced pressure and a plasma, but not evaporation, to generate a flux of the desired composition for coating a substrate. A heavy inert working gas (usually argon) is leaked into a vacuum chamber to provide a partial pressure of about 1-100 mTorr. This gas is ionized by imposing a voltage on the order of 500-5000 V, with the substrate charged positive with respect to a metallic target. The target consists of the elements intended for deposition as the coating. The glow discharge from such a diode arrangement results from the ionization of the heavy gas, and the resulting ions are accelerated at high energy to impact with the target. These collisions knock out atoms, molecules, and clusters from the target, and these species with high kinetic energies of 10-40 eV condense to form the coating. A triode arrangement can also be used to generate the plasma independently of the target. Direct-current voltages are generally applied, but radio frequency potentials must be applied to overcome charge accumulation for insulating targets. Because of the high ionization efficiency in the magnetron cavity, intense plasma discharges that provide high sputtering rates can be maintained at moderate voltages, even for low pressures.

The primary advantage of sputtering is its ability to deposit a wide variety of materials (e.g., alloys, oxide solutions, and intermetallics). These compositions can be derived from targets of many types: segments of several materials, several different targets used simultaneously, several targets used sequentially to deposit a laminated coating, etc. While different materials exhibit differing sputtering yields (defined as the number of target atoms ejected per incident particle), these factors are known in the technology (Bunshah, 1982). In fact, the sputtering yields for different materials depend on the nature of the bonding in the materials and are a function of the incident ion energy. These yields differ significantly for quite different materials, somewhat analogously to the differing vapor pressures of components in a liquid evaporation source.

A reactive gas can be introduced with the heavy inert gas, so that reactive sputtering results, again, from reactive collisions in the gas phase. Although the substrate can, in principle, be heated to any temperature, it is usually held at a temperature that is too low to achieve significant interdiffusion during deposition. After cooling from the processing temperature and heating to a higher service temperature, the inherent differences in the expansion coefficients between the substrate and the coating generate residual stresses of opposing sign. Often the deposit achieves a columnar microstructure, with elongated grains normal to the interface. As previously mentioned, such a structure is ideal for oxide TBCs.

Sputtering is currently not a production coating method for turbine hardware because of the slow deposition rate of current equipment. However, the method does provide a pure coating of nearly any desired composition.

ELECTROSPARK DEPOSITION

Electrospark deposition is a microwelding process that uses pulsed electrical arcs to deposit an electrode material onto a metallic substrate (Johnson, 1995). The process yields functionally graded coatings (with a gradient dependent on deposition rate) by depositing material in a series of passes. Substrate materials must be electrically conductive and capable of being melted. Coatable surface geometry is limited by access to the surface by the electrode. Electrospark deposition has mainly been used as a repair technique for gas-turbine engine components, although electrospark deposition has been used to apply platinum as the first step in some platinum-modified aluminide diffusion coating processes. The metallurgical bonding of electrospark deposition has been demonstrated in tests to be more resistant to spalling than mechanically bonded coatings (Johnson, 1995).

AUXILIARY PROCESSES

Auxiliary equipment needed for coating production using any of the above processes may also rely on chemical cleaning (vapor degreaser, ultrasonic cleaning, and alkaline cleaning tanks), surface prep (wet or dry grit blasting), mechanical and chemical masking for protecting noncoated areas, and vacuum heat-treatment furnaces for post-coating heat treatment.

QUALITY CONTROL

Metallurgical cross-sectioning is still the primary method used to determine conformance to overall quality requirements, such as thickness, coating structure, interface quality, and masking transitions. Witness samples used for metallurgical evaluations are designed to be representative of the coating batch or lot with which they were processed and are generally scrap components. The trend in recent years has been for much tighter control of the coating process because many components cannot be stripped and recoated if the coating is nonconforming. This has led to the use of process capability studies prior to committing to production process qualifications, the use of statistical process control once in production to assess manufacturing process variation, and the use of Taguchi and integrated process management to make the process more robust (Slater, 1991; Taguchi, 1993). The process capability ratio, C_{pk}, is a simple but powerful measure of the ability of a manufacturing process to meet specification requirements and is defined as C_{pk} = (6 standard deviation)/(specification requirement). Ratios less than 75 percent are considered to be capable of running without significant risk of nonconformance and ratios less than 30 percent are at world-class level. Process capability studies are not yet a part of the quality standards used by coatings suppliers, although the process is beginning to be used for control of other manufacturing operations.

The specific trends in coating manufacturing can also be examined in terms of engine class. The small (turboprop) engines have moved to directionally solidified and single-crystal high-pressure turbine blades and vanes, which generally use serpentine internal cooling passages and trailing-edge and tip-cooling holes. Film cooling is not common, however, and these designs do not contain external cooling holes. Because of the small size and thin-wall thicknesses of the components, the possibility of stripping and recoating is limited, so coating thickness must be closely controlled.

Aluminide and platinum-modified aluminide coatings are widely used (pack or CVD if machined after coating). Limited MCrAlY (usually EB-PVD) and some EB-PVD TBCs are now in use.

The medium-size gas-turbine engines typified by medium-range commercial aircraft propulsion are making extensive use of directionally solidified and single-crystal high-pressure turbine blades and vanes. Serpentine internal passages and complex film-cooling schemes are commonplace. Due to the criticality of air flow requirements, the control and shape of hole diameter is essential. Often the hole diameter must be properly sized to the coating process, and changes in the process (e.g., pack to CVD or the addition of a TBC) necessitate a change in the targeted hole diameter. Coating process variability can also have a large impact on meeting air flow requirements. Strip and recoat is generally restricted to one time because of wall thickness concerns. There is widespread use of MCrAlY (LPPS) and platinum-modified aluminide (CVD) coatings in combination with internal coatings (CVD and coatings from the Advanced Technology Program). EB-PVD TBCs are widely used in conjunction with the above internal and external coatings.

The trends for large turbines (land-based power generation) have paralleled those for the aircraft engines. Design engineers are moving to directionally solidified and single-crystal high-pressure turbine blades for the newest engines. Serpentine cooling is also being considered. The platinum-modified aluminide coating is generally being replaced by thick MCrAlYs (10-15 mils) produced by LPPS. In some cases the MCrAlY is overaluminided. Gas-phase coatings and TBCs (plasma and EB-PVD) are being evaluated; strip and recoat is common. Internal coatings are provided by pack and slurry methods.

REFERENCES

Benden, R., and R. Parzuchowski. 1979. Apparatus for Gas Phase Deposition of Coatings. U.S. Patent, Number 4,148,275.

Bianco, R., and R.A. Rapp. 1993. Pack cementation alumide coatings on superalloys. Codeposition of Cr and reactive elements. Journal of the Electrochemical Society 140(4):1181-1191.

Bunshah, R.F., ed. 1982. Deposition Technologies for Films and Coatings. Park Ridge, New Jersey: Noyes.

Chraska, P., and M. Hrabovsky. 1992. An overview of water stabilized plasma guns and their applications. Pp. 81-85 in Thermal Spray: International Advances in Coatings Technology, C.C. Berndt, ed. Materials Park, Ohio: ASM International.

Gauje, R., and R. Morbioli. 1982. Vapor-phase aluminizing to protect turbine airfoils. Journal of Metals 35(12):A12.

Goward, G.W., and D.H. Boone. 1971. Mechanisms of formation of diffusion on aluminide coatings on nickel-base superalloys. Oxidation of Metals 3(5):475.

Goward, G.W., and L.W. Cannon. 1988. Pack Cementation Coatings for SuperalloysHistory, Theory and Practice. ASME Paper 87-GT-50. New York: American Society of Mechanical Engineers.

Herman, H. 1991. Powders for thermal spray technology. Thermal Spray Technology. Powder Science and Technology 9:187-199.

Johnson, R.N. 1995. Electrospark deposited coatings for high temperature wear and corrosion applications. Pp. 265-277 in Elevated Temperature Coatings: Science and Technology I, N.B. Dahotre, J.M. Hampikian, and J.J. Stiglich, eds. Warrendale, Pennsylvania.: TMS.

Kaufold, R.W., A.J. Rotolico, J. Nerz, and B.A. Kushner. 1990. Deposition of coatings using a new high velocity combustion spray gun. Pp. 561-569 in Thermal Spray Research and Applications, T.F. Bernecki, ed. Materials Park, Ohio: ASM International.

Levine, S.R., and R.M. Caves. 1973. Thermodynamics and kinetics of pack aluminide coating formation. Journal of the Electrochemical Society 120(8):C232.

Muehlberger, E. 1973 A high energy plasma coating process. In the Seventh International Metal Spraying Conference, London, September 10-14. Cambridge, England: British Welding Institute.

Russo, L., and M. Dorfman. 1995. High-temperature oxidation of MCrAlY coatings produced by HVOF. Pp. 1179-1184 in Proceedings of the International Thermal Spray Conference, A. Ohmori, ed. Japan: High-Temperature Society of Japan.

Slater, R. 1991. Integrated Process Management: A Quality Model. New York: McGraw Hill.

Smith, J.S., and D.H. Boone. 1990. Platinum-Modified Aluminides-Present Status. Paper presented at the International Gas Turbine and Aeroengine Congress and Exposition, Brussels, Belgium, June.

Taguchi, G. 1993. Robust Development. New York: American Society of Mechanical Engineers Press.

Appendix F

Example of a Coating Designation System

Any standard designation system for coatings must capture the essential features of the materials and processes being used and be sufficiently flexible so that future materials and processes can readily be added without requiring a new framework. A potential designation system would also have to be agreed on by all interested parties before it could be adopted. The purpose of this appendix is to suggest some possible approaches to such a system.

An example of a standard coating designation system is a code with four elements. The code would be "XX####-X###-T##(M or E)." The code has the flexibility to be used by the first part alone, the first two parts, the first three parts, or the entire code. The coating is progressively better defined as more parts of the code are used.

The first part of the code, "XX," refers to the class of coating, such as AL for a simple aluminide, AS for a silicon modified aluminide, AP for a platinum aluminide, CR for a chromide, NI for a nickel-base overlay, CO for a cobalt-base overlay, and NC for nickel/cobalt-base overlay.

The second part of the code, "####," is four numbers that give the nominal chemical composition. For a simple aluminide coating, this would be the aluminum level in weight percent, such as "2300" for 23-weight-percent aluminum. For a platinum aluminide this would be the aluminum level followed by the platinum level, such as "2310" for 23-weight-percent aluminum and 10-weight-percent platinum. For an overlay coating, the numbers would be the chromium and aluminum levels in weight percent, such as "2308" for 23-weight-percent chromium and 8-weight-percent aluminum.

The third part of the code, "X###," refers to processing, chemistry, or base-metal requirements. They are defined by reference to a table of terms, such as for the aluminum standards. Processing codes would start with a "P" and would refer to a slurry process, an inward diffusion coating, an air plasma spray, a low-pressure plasma spray, etc. Chemistry codes would start with a "C" and refer to special alloying elements, such as hafnium or silicon. Base-metal codes would start with a "B" and would refer to the specific superalloy types for which the coating is intended. Multiple designators would be used as needed. This portion of the code has the greatest potential for creating a very long code.

The last part of the code, "T##(M or E)," is the nominal thickness of the coating, expressed in either millimeters or mils. The "M" is used for millimeters and the "E" is used for mils to identify the units of thickness.

Duplex coatings and thermal barrier coatings (TBCs) require that the individual parts of the coating be separately specified. Thus a duplex coating would be called "DPXX/YY" and a TBC would be "TB-XX/YY," where XX would be the top coating and YY the bottom coating. As an example, an aluminide on top of a NiCrAlY would be called "DP-AL/NI." A TBC over the same bondcoat would be "TB-ZR/NI." The codes for the component coatings would still be required after this designation.

The advantage of this coding system is that it captures the essential features of the coating. The disadvantage is its potential for a long and ungainly designator.

Alternatives to this system are to use either a number for each coating, similar to what is currently used by vendors, or a two-digit letter to identify the class of coating (e.g., AL, AP, CO) followed by a number for the specific type of coating. Both of these alternatives have the advantage of a simpler designator but the disadvantage of providing little information about the coating from the designator.

Appendix G

Biographical Sketches of Committee Members

ROBERT V. HILLERY, *chair,* is manager of Airfoil Materials in the Engineering Materials Technology Laboratories at GE Aircraft Engines, Cincinnati, Ohio. He has a B.Sc. degree in metallurgy and a Ph.D. degree in corrosion science from the University of Manchester. His research interests are in environmental high-temperature oxidation, corrosion-resistant thermal barrier wear, and erosion coatings and their processing.

NEIL BARTLETT is professor of chemistry at the University of California, Berkeley. He has B.Sc. and Ph.D. degrees in inorganic chemistry from the University of Durham. He is a foreign associate of the National Academy of Sciences and a member of the Royal Society of London, the Gottingen Academy of Sciences, the American Academy of Arts and Science, and the Academy of Science of France. His research concerns fluorine inorganic chemistry, noble gas chemistry, high-energy oxidizers, X-ray crystallography, solid-state chemistry, nonaqueous solvent chemistry, and thermochemistry.

HENRY L. BERNSTEIN is a staff engineer in the Materials and Structures Division and the assistant director of the Electric Power Research Institute's Materials Center for Combustion Turbines at Southwest Research Institute. He is a fellow of the American Society of Mechanical Engineers. He has a B.S. degree from Ohio State University, an M.S. degree from the University of Illinois at Urbana, and a Ph.D. degree in applied mechanics and materials science from the University of Cincinnati. His research concerns gas-turbine durability and life prediction and the mechanical behavior of high-temperature materials. In both of these areas, high-temperature coatings are of key interest.

ROBERT F. DAVIS is a professor of materials science and engineering at North Carolina State University. He has a B.S. degree from North Carolina State University, an M.S. degree from Pennsylvania State University, and a Ph.D. degree in ceramic engineering from the University of California, Berkeley. His research interests are in diffusion and high-temperature deformation of ceramic materials, growth and characterization of wide bandgap electronic materials, and ceramic/metal coating deposition and interface adherence.

HERBERT HERMAN is Leading Professor of materials science at the State University of New York at Stony Brook. He has a B.S. degree from De Paul University and M.S. and Ph.D. degrees in metallurgy from Northwestern University. His research interests are in thermal-sprayed protective coatings, intermetallic compounds, metal-and ceramic-matrix composites, fuel cell processing, powder metallurgy and processing, and corrosion protection of transportation and marine structures.

LULU L. HSU is chief of process engineering at Solar Turbines in San Diego, California. She has a B.A. degree from Barat College and an M.S. degree in analytical chemistry from Michigan State University. She has 19 years of experience in the gas-turbine industry with Solar Turbines Incorporated. For 17 of these years she did work in the field of materials engineering, where she was involved in the development, evaluation, and application of hot-section coatings for use in Solar's fleet of industrial gas turbines. During this period, she collaborated with other turbomachinery experts and evolved a comprehensive approach on contaminant ingestion within the context of air, fuel, and water management. Over the years, she was also been involved with providing technical support on engine cleaning, lube oils for turbomachinery, and component repairs, including responsibility for Solar's engineering specifications on the same subjects. In her current role as customer services marketing manager she provides technical and commercial support to Solar's aftermarket product offerings.

WEN L. HSU is a distinguished member of the technical staff and technical coordinator of ultrahard materials for Sandia National Laboratories in Livermore, California. He has a B.S. degree from Rensselaer Polytechnical Institute and a Ph.D. degree in astrophysical sciences-plasma physics from Princeton University. His research concerns diamond, ceramic materials, plasma chemical vapor deposition, and chemical vapor deposition reaction kinetics.

JOHN C. MURPHY is a professor in the Materials Science and Biomedical Engineering departments and principle staff physicist at the Applied Physics Laboratory at Johns Hopkins University. He has a B.A. degree from the Catholic University of America and M.S. and Ph.D. degrees in physics from the University of Notre Dame. His research interests include microwave-optical double resonance experiments on excitation migration in solids; photocatalysis; elastic wave propagation in soils using electro-optical methods; photoacoustic spectroscopy; electron spin resonance of electrogenerated radical ions in solution; photothermal imaging; thermal and thermoacoustic imaging and spectroscopy; microwave and magnetic properties of solids; metrology; nondestructive evaluation of materials; and near-field imaging of organic conductors and films.

ROBERT A. RAPP is professor emeritus of materials science and engineering at Ohio State University. He has a B.S. degree from Purdue University and M.S. and Ph.D. degrees in metallurgical engineering from the Carnegie Institute of Technology. His research interests include the oxidation, corrosion, and coating of metals and alloys. He is a member of the National Academy of Engineering.

JEFFERY S. SMITH is manager of surface technology at Howmet's Operhall Research Center. He has B.S. and M.S. degrees in metallurgical engineering from the University of Wisconsin, Madison. His research interests are superalloy processing and manufacturing methods, superalloy coating process development and performance, high-temperature oxidation and repair technology for engine-run components. He is currently chair of the Manufacturing, Materials, and Metallurgy Committee of the International Gas Turbine Institute of the American Society of Mechanical Engineers.

JOHN STRINGER is technical director of exploratory research at the Electric Power Research Institute. He has B.E. and Ph.D. degrees from the University of Liverpool. His research concerns high-temperature oxidation and corrosion of metals and alloys, fossil-fuel burning systems, hot-corrosion of gas turbines, and erosion and erosion/corrosion of metals and alloys.